Communications in Computer and Information Science   2396

Series Editors

Gang Liⓘ, *School of Information Technology, Deakin University, Burwood, VIC, Australia*
Joaquim Filipe ⓘ, *Polytechnic Institute of Setúbal, Setúbal, Portugal*
Zhiwei Xu, *Chinese Academy of Sciences, Beijing, China*

Rationale
The CCIS series is devoted to the publication of proceedings of computer science conferences. Its aim is to efficiently disseminate original research results in informatics in printed and electronic form. While the focus is on publication of peer-reviewed full papers presenting mature work, inclusion of reviewed short papers reporting on work in progress is welcome, too. Besides globally relevant meetings with internationally representative program committees guaranteeing a strict peer-reviewing and paper selection process, conferences run by societies or of high regional or national relevance are also considered for publication.

Topics
The topical scope of CCIS spans the entire spectrum of informatics ranging from foundational topics in the theory of computing to information and communications science and technology and a broad variety of interdisciplinary application fields.

Information for Volume Editors and Authors
Publication in CCIS is free of charge. No royalties are paid, however, we offer registered conference participants temporary free access to the online version of the conference proceedings on SpringerLink (http://link.springer.com) by means of an http referrer from the conference website and/or a number of complimentary printed copies, as specified in the official acceptance email of the event.

CCIS proceedings can be published in time for distribution at conferences or as post-proceedings, and delivered in the form of printed books and/or electronically as USBs and/or e-content licenses for accessing proceedings at SpringerLink. Furthermore, CCIS proceedings are included in the CCIS electronic book series hosted in the SpringerLink digital library at http://link.springer.com/bookseries/7899. Conferences publishing in CCIS are allowed to use Online Conference Service (OCS) for managing the whole proceedings lifecycle (from submission and reviewing to preparing for publication) free of charge.

Publication process
The language of publication is exclusively English. Authors publishing in CCIS have to sign the Springer CCIS copyright transfer form, however, they are free to use their material published in CCIS for substantially changed, more elaborate subsequent publications elsewhere. For the preparation of the camera-ready papers/files, authors have to strictly adhere to the Springer CCIS Authors' Instructions and are strongly encouraged to use the CCIS LaTeX style files or templates.

Abstracting/Indexing
CCIS is abstracted/indexed in DBLP, Google Scholar, EI-Compendex, Mathematical Reviews, SCImago, Scopus. CCIS volumes are also submitted for the inclusion in ISI Proceedings.

How to start
To start the evaluation of your proposal for inclusion in the CCIS series, please send an e-mail to ccis@springer.com.

Nurit Haspel · Kevin Molloy
Editors

Computational Structural Bioinformatics

International Workshop, CSBW 2024
Boston, MA, USA, November 16, 2024
Proceedings

Editors
Nurit Haspel
University of Massachusetts Boston
Boston, MA, USA

Kevin Molloy
James Madison University
Harrisonburg, VA, USA

ISSN 1865-0929   ISSN 1865-0937 (electronic)
Communications in Computer and Information Science
ISBN 978-3-031-85434-7   ISBN 978-3-031-85435-4 (eBook)
https://doi.org/10.1007/978-3-031-85435-4

© The Editor(s) (if applicable) and The Author(s), under exclusive license
to Springer Nature Switzerland AG 2025

This work is subject to copyright. All rights are solely and exclusively licensed by the Publisher, whether the whole or part of the material is concerned, specifically the rights of translation, reprinting, reuse of illustrations, recitation, broadcasting, reproduction on microfilms or in any other physical way, and transmission or information storage and retrieval, electronic adaptation, computer software, or by similar or dissimilar methodology now known or hereafter developed.
The use of general descriptive names, registered names, trademarks, service marks, etc. in this publication does not imply, even in the absence of a specific statement, that such names are exempt from the relevant protective laws and regulations and therefore free for general use.
The publisher, the authors and the editors are safe to assume that the advice and information in this book are believed to be true and accurate at the date of publication. Neither the publisher nor the authors or the editors give a warranty, expressed or implied, with respect to the material contained herein or for any errors or omissions that may have been made. The publisher remains neutral with regard to jurisdictional claims in published maps and institutional affiliations.

This Springer imprint is published by the registered company Springer Nature Switzerland AG
The registered company address is: Gewerbestrasse 11, 6330 Cham, Switzerland

Preface

Recent years have seen a rapid accumulation of macromolecular structures, as well as recent advances in the computational modeling and prediction of protein structures, especially using deep learning and large language models. This presents a unique set of challenges and opportunities in the analysis, comparison, modeling, and prediction of macromolecular structures and protein-protein interactions. The 2024 Computational Structural Bioinformatics Workshop (CSBW) was held in Boston on November 16, 2024. The workshop papers explored relevant problems in classification, modeling, and analyzing protein structures and complexes. This issue includes the seven papers presented at the workshop. This is the 17th time CSBW has been held since its inception in 2007. Past workshops were held in conjunction with ACM-BCB or IEEE BIBM. 2024 was the first year the workshop was held independently.

CSBW 2024

The CSBW 2024 was chaired by Nurit Haspel (UMass Boston) and Kevin Molloy (James Madison University). The program committee consisted of ten members. Each of the seven submissions was reviewed three times in a single blind process. The papers were evaluated based on quality, originality, and relevance to the workshop. The reviewing process was supported by the EasyChair conference system.

The seven papers accepted for presentation are outlined below:

In "*SuperFoldAE: Enhancing Protein Fold Classification with Autoencoders*", the authors introduce a 2D-convolutional autoencoder neural network specifically designed for protein fold classification. The model employs advanced convolutional techniques and a supervised loss on the autoencoder's latent layer to effectively learn patterns in protein structures.

In "*Enhancing Drug Discovery via Physics-Guided Deep Generative Models*", the authors test physics-guided deep generative models by generating corresponding drug molecule candidates for a variety of protein-ligand complexes from the PDBBind dataset. On average, more than 75% of the structures generated by a hybrid model proposed by the authors are stronger binders than the original reference ligands to the protein and had better binding affinities.

In "*Decoys Reveal Multiple Basins of Attraction for Cryo-Electron Microscopy Flexible Fitting*", the authors explore the robustness and uniqueness of the flexible fitting of atomic structures against cryo-electron microscopy (cryo-EM) maps using elastic network motion models. They systematically generate decoys using normal modes, offering a broader sampling of the conformational space compared to a single-start structure. This strategy allows exploration of the global properties of the cross-correlation (CC) scoring function landscape.

In "*Using Autoencoders to Explore the Conformational Space of the Cdc42 Protein*", the authors train a neural network model on the MD simulations data to create a low-dimensional latent space. The latent space is further used to explore the protein's conformational space. It can be used to be interpolated or extrapolated to produce new intermediate protein conformations that might not have been previously seen.

In *"Using Molecular Dynamics to Assess How Two Insertion Mutations Affect Protein Structure"*, the authors explore the use of molecular dynamics (MD) to assess the effects of pairs of insertion mutations into the PDB structure file of HIV-1 Protease. They use an in-house compute pipeline to generate the exhaustive set of mutants with two insertion mutations, and identify 24 mutant structures which present as outliers. The MD analysis can reveal the effects of the insertion mutations that earlier work was unable to elucidate.

The paper *"Predicted Multimers using Alphafold-3 and Detected Secondary Structure Elements from Cryo-EM maps - Case Studies"* involves four AF3 multimer models and corresponding cryo-EM maps with 7–8 Å resolution. The results show that the predicted multimer models are partially correct. The predicted models contain fairly accurate domains, secondary structures, and individual chains. The analysis illustrates the potential for improvements in the accuracy of AF3-predicted multimer models by combining the density map–model similarity (CC scores) and the secondary structure map–model similarity in a future approach.

In *"Automatically Explaining Binding Mechanisms: A Preliminary Study"*, the authors present a novel approach to interpreting mechanistic insights from machine learning methods. They manually annotated a dataset of 1,225 mutation experiments with mechanistic insights focused on electrostatic, hydrogen bonding, steric, and hydrophobic interactions. Using a Gradient-Boosting Trees (GBT) model trained to predict binding affinity, they authors find that the SHapley Additive exPlanations (SHAP) values generally agree with the annotated mechanisms from the dataset, especially when looking at electrostatic and steric features. They also find that hydrophobicity consistently plays a dominant role and hydrogen bonds consistently play a secondary role, challenging conventional assumptions about the role of these interactions.

November 2024

Nurit Haspel
Kevin Molloy

Organization

Program Chairs

Nurit Haspel — University of Massachusetts Boston, USA
Kevin Molloy — James Madison University, USA

Program Committee

Filip Jagodzinski — Western Washington University, USA
Gideon Gogovi — George Mason University, USA
Nasrin Akhter — University at Buffalo, USA
Toki Tahmid Inan — George Mason University, USA
Jing He — Old Dominion University, USA
Hyuntae Na — Pennsylvania State University, USA
Bruna Jacobson — University of New Mexico, USA
Yonggang Lu — Lanzhou University, China
Negin Forouzesh — California State University, Los Angeles, USA
Brian Chen — Lehigh University, USA

Contents

SuperFoldAE: Enhancing Protein Fold Classification with Autoencoders 1
   Shraddha Patre, Riya Kanani, and Fardina Fathmiul Alam

Enhancing Drug Discovery via Physics-Guided Deep Generative Models 16
   Dikshant Sagar, Ari Jasko, and Negin Forouzesh

Decoys Reveal Multiple Basins of Attraction for Cryo-Electron-Microscopy
Flexible Fitting 31
   Maytha Alshammari, Jing He, and Willy Wriggers

Using Autoencoders to Explore the Conformational Space of the Cdc42
Protein 45
   Fatemeh Afrasiabi, Ramin Dehghanpoor, and Nurit Haspel

Using Molecular Dynamics to Assess How Two Insertion Mutations Affect
Protein Structure 58
   Changrui Li, Katie Christensen, Sarah Coffland, and Filip Jagodzinski

Toward Modeling Protein Multimers by Combining AlphaFold 3 Predictions
with Secondary Structures from Medium-Resolution Cryo-EM Maps 71
   Changrui Li, Thu Nguyen, Willy Wriggers, and Jing He

Automatic Explanation of Protein-Protein Binding Mechanism:
A Preliminary Study 84
   *Justin Z. Tam, Yangying Liu, Dhruv S. Jain, Grant Armstrong,
   and Brian Y. Chen*

Author Index 99

SuperFoldAE: Enhancing Protein Fold Classification with Autoencoders

Shraddha Patre, Riya Kanani, and Fardina Fathmiul Alam(✉)

Department of Computer Science, University of Maryland, College Park, MD 20742, USA
{spatre,rkanani}@terpmail.umd.edu, fardina@umd.edu

Abstract. Protein fold classification is essential for understanding protein function and its role in biological processes. Proteins are crucial for cellular structure, organization, and function, with their tertiary structure directly influencing their roles within cells. This drives interest in computational research, particularly machine learning approaches for classifying protein folds from sequence data. While numerous methods exist for classifying protein folds from sequence data using machine learning, few approaches leverage tertiary structure data with deep learning techniques. In this paper, we introduce SuperFoldAE, a 2D-convolutional autoencoder neural network specifically designed for protein fold classification. Our model employs advanced convolutional techniques and a supervised loss on the autoencoder's latent layer to effectively learn patterns in protein structures. Autoencoders excel at compressing high-dimensional data into a lower-dimensional latent space, capturing intricate patterns and critical features of protein structures. We investigate various configurations within a supervised framework and enhance generalization by integrating unsupervised regularizers via reconstruction loss. Our approach aims to improve classification accuracy by learning representations that highlight subtle structural variations often overlooked in sequence data alone. Using the Structural Classification of Proteins (SCOP) 1.75 dataset, we seek to enhance the reliability of protein fold classification. Our findings contribute valuable insights into the functional implications of protein structures and pave the way for further research into representation-based deep learning for protein classification.

Keywords: protein modeling · tertiary structure · fold classification · representation learning · autoencoder · machine learning · neural network · latent representation · unsupervised regularizers · generalization

1 Introduction

Protein fold recognition is a fundamental task in structural biology, essential for understanding the functional roles of proteins and their evolutionary relationships. As the volume of high-resolution structural data in repositories like the Protein Data Bank (PDB) [6] expands, there is a growing demand for fast, accurate, and fully automated classification methods.

Despite significant progress in recent years, accurately classifying protein structures remains challenging due to the complex relationships between amino acid sequences and their resulting three-dimensional (3D) structures [9]. Traditionally, fold recognition has relied on two primary methodologies: sequence alignment-based methods [11,15] and machine learning techniques [21]. Sequence alignment methods compare a query protein sequence with known structural templates to identify the most likely fold. Early machine learning approaches, notably support vector machines (SVMs) and ensemble classifiers, utilize sequence correlation-based features to enhance fold recognition. While SVMs effectively classify closely related protein families, their performance can falter with highly dissimilar sequences or insufficient training data. Ensemble classifiers improve accuracy by combining multiple models, reducing overfitting, and enhancing generalization.

However, fold recognition becomes significantly more complex with sequences that lack clear evolutionary relationships, highlighting the need for incorporating structural features into predictive frameworks. Information from tertiary structures allows us to capture intricate interactions and patterns not easily discernible through sequence analysis, thereby improving fold classification accuracy. Despite advancements, challenges persist due to the vast number of potential structures and high data dimensionality. Standard machine learning models like SVMs and multi-layer perceptrons (MLPs) often struggle with thousands of protein classes [2]. Although recent deep learning advancements - particularly convolutional neural networks (CNNs) [12] and recurrent neural networks (RNNs) - show promise for feature extraction, they still face overfitting challenges [1]. To address this, Supervised Autoencoders (SAEs) - supervised learning techniques that make use of unsupervised regularizers [13] - offer a promising approach. These regularizers help control model complexity and improve generalization by minimizing reconstruction error.

Inspired by this work in Computer Vision, we propose a 2D Autoencoder Convolutional Neural Network model, SuperFoldAE. Our approach utilizes SAEs to classify tertiary protein structures. By combining dimensionality reduction through autoencoders with a supervised learning component, our model reduces the complexity of tertiary protein data while ensuring relevant representations for both reconstruction and classification tasks. We evaluate whether the SuperFoldAE architecture, leveraging learned representations, surpasses traditional models in classifying protein folds and improves generalization on unseen data. Using the SCOP 1.75 dataset, we compare our model's performance against other baselines and AE-Classification models, analyzing the effect of unsupervised regularization on classification accuracy. This approach aims to provide a reliable framework for classifying tertiary protein structures, contributing to deeper insights into protein functionality.

2 Related Works

Driven by the rapid advancements in neural network research, we explore various autoencoder (AE) architectures for effectively reducing the dimensionality of protein structure data. Our investigation is primarily focused on utilizing AEs for learning meaningful representations, which can later be leveraged for protein fold classification. The concept of autoencoders was first introduced in [7], where an AE was applied to a substituted cyclo-octane consisting of 24 atoms. Later some works, such as [14], have employed AEs with multiple hidden layers in the encoder alongside innovative cost functions to analyze molecular dynamics simulation data, specifically for Asp7–a small molecule characterized by 12 backbone dihedral angles–and Trp-cage, a compact protein comprising 20 amino acids. Additionally, research in [8] examined a similar AE architecture to summarize the folding landscape of Trp-cage. Our approach builds on these foundations, focusing on the representation learning capabilities of AEs to enhance the subsequent classification of protein folds.

Following these foundational works, [2,3] investigated various shallow and deep AE architectures to reduce the dimensionality of protein structure data generated by template-free protein structure prediction methods while preserving essential structural features. Their application of autoencoder-based representations addresses the classic problem of decoy selection [3] in protein structure prediction. By employing off-the-shelf supervised learning methods, they demonstrated that these learned features are meaningful and effective in identifying active tertiary structures. The methodology highlighted in [2,3] underscores the potential of autoencoders not only in reducing data dimensionality but also in maintaining high performance for complex biological datasets like protein tertiary structures. This paves the way for exploration of the applications of representation learning techniques to other tasks, such as protein fold classification.

The work by [13] analyzes how adding reconstruction error in supervised autoencoders (SAEs) can enhance generalization. It shows that for linear autoencoders, reconstruction error acts as a regularizer, promoting stability without the negative bias of traditional methods like L2 norms. Their research suggests that jointly learning multiple tasks enhances generalization and stability by treating reconstruction as an auxiliary task, thereby improving model robustness without compromising accuracy. This finding is crucial in fields like protein structure prediction, where effective generalization to unseen data is vital.

As detailed in Sect. 1, this paper aims to advance the research on utilizing learned latent representations of tertiary structures for protein fold classification. Unlike previous works, our proposed deep convolutional autoencoder-based neural network distinguishes itself by directly learning these representations from experimentally available tertiary structures. This learned representation is then effectively integrated into the subsequent CNN classification phase of the model, enhancing the overall accuracy of fold classification. By leveraging distance matrix representations, our approach is specifically designed to capture intricate structural features that are crucial for accurate classification. We will elaborate on our methodological approach in Sect. 3.

3 Methods

3.1 Training Dataset and Tertiary Structure Representation

In this paper, we utilized the SCOP 1.75 dataset, as detailed in [12]. The Structural Classification of Proteins (SCOP) [16] provides a standardized taxonomic classification of protein domain structures based on evolutionary relationships, emphasizing connections among homologous sequences to enhance our understanding of protein function. The SCOP 1.75 dataset, released in 2009, includes genetic domain sequence subsets with less than 95% pairwise identity, comprising 16,712 proteins classified into seven major structural classes and 1,195 identified folds. We obtained the experimentally determined tertiary structures of these proteins from the Protein Data Bank (PDB) [6].

To ensure consistency and comparability with the CNN model in [12], we adhered to the same distribution of training and testing data. Specifically, we employed the protein list utilized in [12], which partitions the dataset into a training set (88%) and a testing set (12%). This partitioning was conducted with the stipulation that no two proteins from the same superfamily were included in both the fold-level training and testing datasets [12].

To train (and test) our model, we employed the distance matrix representation of the tertiary protein structures [4,5,19]. This representation captures the spatial relationships among amino acids. Since our dataset contains structural information for proteins across 1,195 different folds, each input consists of 2D distance maps derived from the central carbon atoms in the original protein structures. Each matrix is a 2D array that represents the distances between the central carbon atoms in the protein. It follows that an amino acid identified to have N central carbon atoms will generate a distance matrix of size $N \times N$.

3.2 Experimental Setup

Given the wide variation in protein lengths–ranging from 9 to 1,419 residues, with most proteins between 9 and 600 residues–we encountered challenges in effectively learning key representations of the data using the original dataset. Our initial experiments indicated that the autoencoder (AE) model struggled to capture essential local and distal patterns present in the protein backbone, as well as short-range and long-range contacts, in the case of smaller-sized distance maps. To tackle the challenges associated with short protein sequences, we standardized the distance matrices. Each protein's distance matrix is generated based on the pairwise distances between central carbon atoms. To ensure uniformity across our dataset, we utilized functions from the Python Imaging Library (PIL) to resize these matrices. The resizing process involves using interpolation techniques to adjust the values in the distance matrix, allowing us to experiment with four size parameters: 64, 128, 256, and 512 [5]. After evaluating their impact on model performance, we ultimately settled on a size of 256, as it provided a balance between capturing sufficient detail and maintaining computational efficiency, optimizing the model's ability to learn relevant representations from the dataset.

During the classification phase of our model, the training data was further divided into training and validation subsets in an 80-20 split. This dataset faces a class imbalance issue, as around 5% of folds contain over 50 instances, while about 69% contain less than 6 instances [12]. Notably, the fold with the highest number of data points was observed to be "b.1," a specific fold in the SCOP classification system with 1,264 instances in the train set. We initially attempted classification across all 1,195 folds but ultimately decided to focus exclusively on the 27 folds with the highest number of instances in the train set to address the class imbalance issue more effectively [20].

We also incorporated data augmentation techniques using Pytorch [17] libraries when training our models. As the classification models from initial implementations, as well as SuperFoldAE, were both trained and tested on only the top 27 classes [20], it was necessary to increase the number of data points to improve classification accuracy and loss. Adding augmented data points to our training and validation datasets increased fold prediction accuracy by almost 4%. Each data point in the training and validation sets was used to create three augmented data points. We separately incorporated a scaling factor of 0.1, a shift factor of 0.1, and a noise factor of 0.05 to generate these augmented data points.

3.3 Autoencoder-Classification Architecture

In our initial implementation, we conducted a preliminary study investigating a multi-layer convolutional autoencoder. The learned representations from the autoencoder are subsequently utilized for fold classification using a separate classification model. Autoencoders (AEs) are neural networks designed to learn efficient data representations in an unsupervised manner. They utilize a deterministic mapping f_θ, parameterized by a vector of parameters $\theta = [W, b]$. This mapping transforms an input vector $\mathbf{x}$ into a hidden representation or code vector $\mathbf{y}$ such that $\mathbf{y} = f_\theta(\mathbf{x})$. The objective of the learning process is to optimize the representation $\mathbf{y}$ by minimizing the reconstruction error, defined as $\mathcal{L}(\mathbf{x}, \hat{\mathbf{x}}) = ||\mathbf{x} - \hat{\mathbf{x}}||^2$, where $\hat{\mathbf{x}}$ is the reconstruction of $\mathbf{x}$. This reconstruction is achieved through an inverse mapping g_θ, expressed as $\hat{\mathbf{x}} = g_\theta(\mathbf{y})$. The ability of AEs to extract key features from high-dimensional data makes them particularly well-suited for complex tasks such as dimensionality reduction and feature extraction. Furthermore, the learned representations $\mathbf{y}$ can be effectively utilized in supervised tasks, such as classification. In such cases, the features obtained from the autoencoder serve as input to a classifier h, which predicts the class label $\mathbf{c}$ as $\mathbf{c} = h(\mathbf{y})$. This integration allows the model to leverage the compact, informative representations for improved performance in classification tasks.

In our autoencoder (AE) model, the encoder comprises two convolutional layers. The first layer applies 16 filters of size 3 × 3 with ReLU activation, followed by batch normalization and max-pooling with a pool size of 2 × 2. The second layer employs 32 filters, also followed by the same activation, batch normalization, and max-pooling. The output from the encoder is then flattened to create the latent representation **z**, which captures the essential features of the protein data in a lower-dimensional space.

The decoder reshapes the latent representation **z** for transposed convolutional operations. It consists of two transposed convolutional layers: the first with 32 filters and the second with 16 filters, each followed by batch normalization and upsampling layers, all aimed at reconstructing the original input **x**. The final output layer employs a single filter with a sigmoid activation function, yielding the reconstructed data $\hat{\mathbf{x}}$. The reconstruction loss, which the model aims to minimize, is calculated using the mean squared error (MSE) formula:

$$\mathcal{L}_{MSE}(\mathbf{x}, \hat{\mathbf{x}}) = \frac{1}{N} \sum_{i=1}^{N} ||\mathbf{x}_i - \hat{\mathbf{x}}_i||^2$$

where N is the number of samples. After training the AE on the entire dataset, which includes all 1,195 folds, we utilize the encoder to generate latent representations **Z** of the filtered training data, focusing on the top 27 classes. We then train a classification model using these latent representations, classifying them into one of 27 classes.

We test this process with multiple classifiers from the scikit-learn library [18], including Random Forests, Logistic Regression, Support Vector Machines (SVM), and the Adaboost ensemble classifier. The Random Forest model performed optimally with 1200 estimators, a maximum depth of 30, and a minimum split of 10 samples. The Logistic Regression model achieved the best performance using the 'lbfgs' solver with an $L2$ penalty and a maximum of 4000 iterations. The SVM produced the best results with a polynomial kernel, using a gamma value of 0.01 and a regularization parameter, $C = 0.1$. Finally, the AdaBoost classifier excelled with a learning rate of 0.01 and a base estimator configured with the parameters of the best-performing Random Forest model. We found that the SVM classifier was able to classify the latent representations better than the other tested classification models.

3.4 SuperFoldAE: Supervised Fold Autoencoder Architecture

To enhance the performance of the AE-classification models, we focus on developing a unified framework that integrates both representation learning and classification within the same model. Our objective is to establish a single model capable of simultaneously learning representations and performing classification, thereby improving overall accuracy and efficiency in recognizing complex patterns within the data. To achieve this, we implement a supervised autoencoder-based classification model (SuperFoldAE) as shown in Fig. 1, which introduces a supervised loss at the latent layer. This supervised component incorporates a classification task, training on the latent representation to predict the protein fold class. This design is inspired by the work in [13], which discusses a supervised AE model created for generalized image classification and tested on the CIFAR10 and SUSY datasets. This foundational research informed our decisions regarding the optimal network size.

The network learns by minimizing a cross-entropy loss function, which measures the difference between the predicted and actual class labels. The cross-entropy loss is defined as:

$$\mathcal{L}_{CE} = -\frac{1}{N}\sum_{i=1}^{N}\sum_{c=1}^{C} y_{i,c} \log(\hat{y}_{i,c}),$$

where N is the number of samples, C is the number of classes, and y is a binary indicator (0 or 1) of whether class c is the correct classification for sample i. Here, $\hat{y}$ refers to the model's prediction. The cross-entropy loss function, commonly used for multi-class classification problems, is crucial in preventing overfitting, particularly in over-parameterized models like deep neural networks.

The SuperFoldAE model deviates from the model discussed in [13] by utilizing mean squared error (MSE) loss (similar to the AE-Classifier models) to calculate the reconstruction error, which helps the network learn to reconstruct the 2D distance maps – a proxy task for learning more accurate representations of the training data. Rather than assigning a larger weight to the classification loss, we improve upon the model by giving equal weight to both classification and reconstruction losses. This balance is crucial as both tasks are equally important in the fold classification problem. The final loss propagated through the network at each epoch is expressed as:

$$\mathcal{L} = \mathcal{L}_{CE} + \mathcal{L}_{MSE},$$

where $\mathcal{L}_{CE}$ and $\mathcal{L}_{MSE}$ are the cross-entropy loss and MSE loss, respectively. Additionally, our model employs a greater number of features to accommodate the larger size of our data points, facilitating the learning of more intricate patterns. We also utilize a Leaky ReLU activation function instead of ReLU to ensure that gradients do not vanish during backpropagation.

Figure 1 illustrates the encoder of the SuperFoldAE, comprised of two convolutional layers with 32 and 64 filters, respectively. These convolutional layers use a kernel size of 3×3, a stride of 1×1, and padding of 1 to extract low-level features. After flattening the output of these convolutional layers, we employ a fully connected layer with 256 output features that form the latent representation. This latent representation is also used in the decoder, which mirrors the architecture of the encoder. The decoder comprises a fully connected layer with 256 input features that transforms the data for use by convolutional layers. The transformed representations feed into two transposed convolutional layers with 64 and 32 filters, with architectures mirroring the layers of the encoder. The output of these convolutional layers is the reconstructed image. Apart from the classification layer, which uses a softmax activation on the latent layer, each layer employs a Leaky ReLU activation with $\alpha = 0.08$.

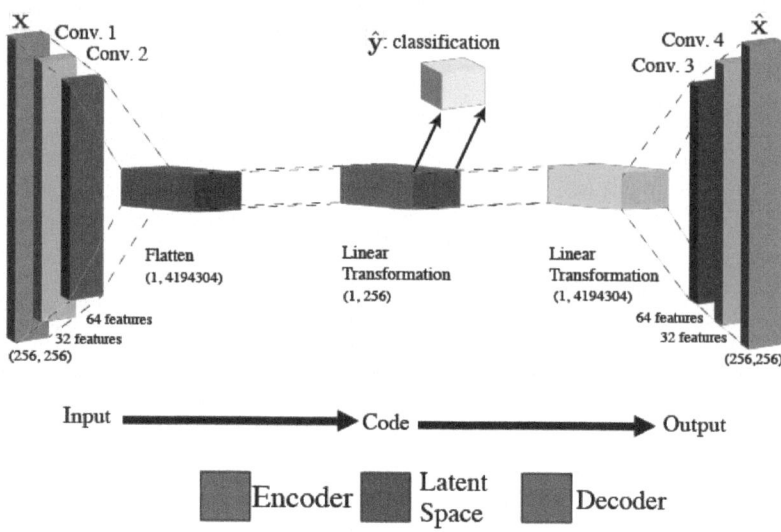

Fig. 1. Architecture of SuperFoldAE. The input layer is a single-channel distance matrix of size $1 \times 256 \times 256$. The 2D convolution layers (Conv.) with 32 and 64 output channels capture complex patterns of the distance maps. The output from the convolutional layers is flattened and processed through a fully connected layer that reduces the dimensionality to 256 units, serving as the encoded representation (latent layer). This representation is transformed to size 1×4194304 using another fully connected layer. A transposed convolution with 64 output channels, followed by a second transposed convolution with 32 output channels, reconstructs to the original image structure. The final Conv. layer with 1 output channel produces the reconstructed output image $\hat{x}$. For the classification output $\hat{y}$, a softmax layer is applied to the latent layer, reducing dimensionality to 27, which indicates the probabilities for each class.

Generalization. One key advantage of SuperFoldAE is its ability to enhance generalization performance through reconstruction loss. We analyze this effect using uniform stability, a property that measures how sensitive a model is to changes in a single training instance. A model is said to be uniformly stable if its predictions do not change significantly when a small portion of the training data is altered. We demonstrate that linear supervised AEs are uniformly stable, which implies better generalization performance. In our case, the reconstruction error acts as an auxiliary task, stabilizing the learning process by preventing the model from overfitting to specific samples. This stability results in a small difference between models trained on different data subsamples, which enhances performance on unseen data. Uniformly stable algorithms are known to exhibit strong generalization performance, and by incorporating reconstruction as a regularizer, the SuperFoldAE model benefits from this property.

3.5 Evaluation and Implementation Details

We implemented, trained, and evaluated the various autoencoder (AE) and classification models using PyTorch [17]. PyTorch is an open-source Python library that provides a high-level interface for deep learning frameworks. In the AE-SVM architecture, the AE was trained for a total of 10 epochs, and a batch size of 64 was utilized. It used the Adam optimizer with a learning rate of 0.01, continuing until convergence was achieved. The SVM classifier uses a polynomial kernel with a gamma value of 0.1 and a C value of 0.1. Figure 2 shows the loss generated by the AE in this model.

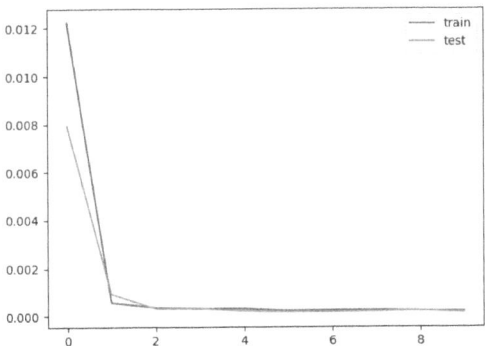

Fig. 2. Loss vs Epochs generated by the autoencoder used in the AE-SVM model

The SuperFoldAE model was trained for 40 epochs, and a batch size of 64 was utilized. The SGD optimizer, with a learning rate of 0.005, momentum of 0.8, and weight decay of 1×10^{-4}, was employed to mitigate the risk of premature convergence. The training times for the SuperFoldAE model ranged from 2,880 to 3,300 s.

In addition to assessing model performance through classification accuracy, we evaluate our models based on precision and recall. We also measure Top 1, Top 5, and Top 10 accuracies to ensure comparability with the baseline DeepSF model [12]. The term *top k accuracy* denotes whether the correct class is included among the top k predicted classes, with Top 1 accuracy aligning with traditional classification accuracy.

4 Results

4.1 Visualization of Distance Matrices

To evaluate the effectiveness of our AE-SVM model in learning representations of experimentally determined tertiary structures, we visualize several distance matrices as heatmaps in Fig. 3 generated by our AE model. We randomly selected three distance maps resized to dimensions 256 × 256 from the SCOP1.75 test dataset. For each protein, we display the original input distance map, along with their latent representations and reconstructed outputs, also sized 256 × 256. We conducted a comparative analysis on the test protein distance maps, the latent representations, and the final reconstructed outputs using the generated heatmaps. This approach enables us to evaluate the AE's capacity to learn meaningful representations of the underlying structural information.

The latent layer consists of 32 feature maps, each with dimensions 64 × 64, resulting in a latent representation of size (32, 64, 64). To visualize the latent layer, we averaged the values across the 32 feature layers for each corresponding point in the matrix, resulting in a 64 × 64 representation. The values in these matrices were normalized, and we employed a yellow-to-blue color scheme to depict the range of distances, with darker areas indicating shorter distances and lighter areas representing longer distances. This visual representation effectively elucidates contact patterns–encompassing both backbone and short-range contacts, as well as long-range interactions–as distinct deep blue lines. Figure 3 demonstrates that the AE models are capable of reconstructing distance matrices that closely resemble those of experimentally determined tertiary structures,

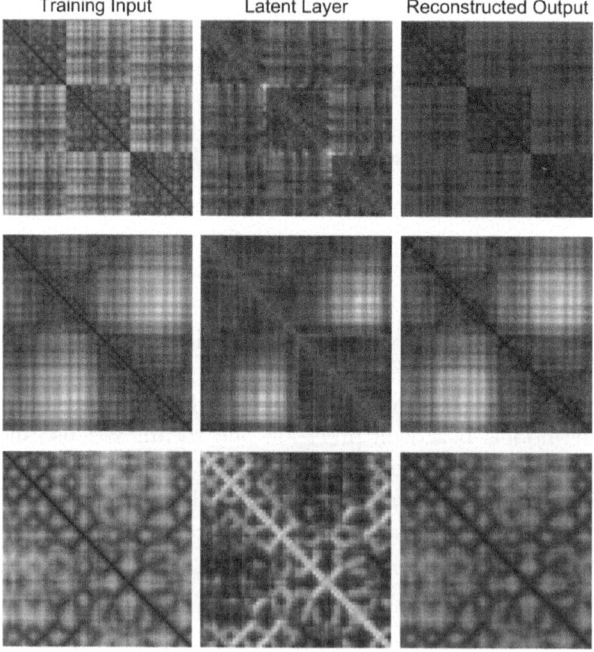

Fig. 3. We visualize three distance matrices from the SCOP1.75 dataset as heatmaps. The left column shows the original 256 × 256 matrices, the middle column presents 64×64 latent representations, and the right column displays the reconstructed matrices. Darker colors indicate shorter distances, while lighter colors represent longer distances.

thereby validating the loss metrics observed during training and confirming their ability to learn meaningful representations. With this foundation established, we shift our focus to the fold classification process, leveraging the learned latent representations.

4.2 Classification Performance on SCOP 1.75 Dataset

Comparative Evaluation of Investigated Models. We first begin by analyzing various AE-classification models alongside SuperFoldAE to identify the optimal model for fold classification. This analysis investigates input distance matrices of varying sizes through four key configurations: 64×64 (representing fragments of 64 amino acids), 128×128, 256×256, and 512×512, generated using the process described earlier in Sect. 3. We compare the classification capabilities of SuperFoldAE against the different AE-classification models in Table 1. This performance evaluation focuses on Top 1 prediction accuracy using the SCOP 1.75 validation dataset. As illustrated in Table 1, the SuperFoldAE model outperforms all other tested models when trained on various matrix sizes, except at the 512×512 size, where its performance is comparable to that of AE-SVM. For the other AE models, which include different classification algorithms such as logistic regression, random forest, SVM, and AdaBoost, we observed that the AE-SVM model performed the best at the 512×512 size. However, SuperFoldAE at 256×256 surpasses these models, achieving the highest prediction accuracy of 88.73%, providing strong evidence for its efficacy. The SVM model with a polynomial kernel outperforms other autoencoder-based classification models by capturing complex nonlinear relationships in the protein distance map while maintaining lower computational demands. Its effective modeling of feature interactions, combined with low γ and C values, enhances generalization. In contrast, SuperFoldAE with unsupervised regularizers achieves superior performance by improving generalization, capturing essential data structures in the distance map, and enhancing downstream classification tasks. In addition to classification accuracy, we also investigate precision and recall metrics to further validate our analysis. We found a validation precision of 87.42% and a recall of 88.13%, as shown in Fig. 4 for SuperFoldAE. Based on these findings, we consider SuperFoldAE$_{256}$ to be our best model and will use it for the remainder of the analysis.

Comparative Evaluation of Fold Level Predictions for Top 1, Top 5, and Top 10 Predictions. Figure 5 presents a comparison of the classification performance of our two highest-performing models, the AE-SVM$_{256}$ model and SuperFoldAE$_{256}$, against the baseline performance of DeepSF and the standard sequence alignment model PSI-BLAST [12]. This comparison extends our analysis to include Top k fold predictions, where k=1,5,10. It is important to note that the accuracy of PSI-BLAST is calculated based on the top folds from the ranked templates [12]. As shown in Fig. 5, SuperFoldAE achieves a performance of 71.75% on the SCOP 1.75 test dataset for Top 1 accuracy at the fold level, significantly higher than PSI-BLAST (5.60%) and DeepSF (40.95%). SuperFoldAE

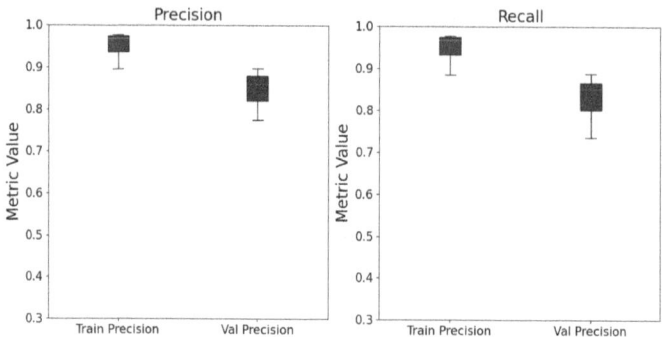

Fig. 4. Box plot of train and validation accuracies over 40 epochs for SuperFoldAE$_{256}$.

Table 1. The fold prediction accuracy of different methods, on SCOP1.75 validation dataset. We investigated input distance matrices of varying sizes by exploring four key configurations: 64×64 (representing fragments of 64 amino acids from the dataset), as well as 128×128, 256×256, and 512×512.

Methods	Fold Prediction Accuracy			
	64×64 input	128×128 input	256×256 input	512×512 input
SuperFoldAE	**73.75%**	**82.12%**	**88.73%**	72.81%
AE-Logistic Regression	41.06%	55.62%	62.60%	72.99%
AE-Random Forest	44.33%	41.87%	50.29%	43.17%
AE-SVM	27.85%	53.11%	72.68%	**75.01%**
AE-Adaboost	41.28%	37.66%	38.55%	38.15%

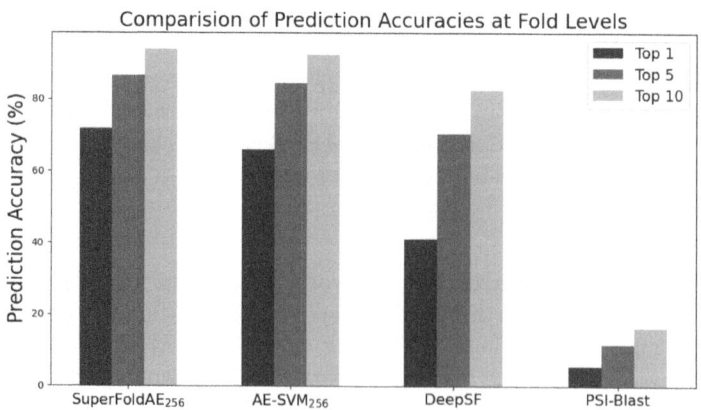

Fig. 5. The prediction accuracy at fold level for top 1, top 5, and top 10 predictions of different methods on the SCOP 1.75 test dataset.

also shows improved performance for Top 5 predictions (86.67%) compared to PSI-BLAST and DeepSF, and excels in Top 10 performance, attaining 93.92%, substantially surpassing both baseline models. Overall, our SuperFoldAE model consistently outperforms all top fold predictions, clearly demonstrating its superior ability to recognize folds compared to PSI-BLAST and DeepSF, even in challenging cases characterized by low sequence identity.

5 Conclusion

In this paper, we presented the SuperFoldAE model, a 2D convolutional autoencoder-based neural network, specifically designed for protein fold classification. Building on previous research that classified 1D protein structure data (DeepSF) [12], we initially employed latent representations generated by a multi-layer convolutional autoencoder as training data for a classification model, achieving notable accuracy on the SCOP 1.75 test dataset with the AE-SVM model. This approach outperformed traditional classifiers such as Logistic Regression, Random Forest, and Adaboost.

To further enhance our implementation, we integrated classification and reconstruction tasks within a single model, resulting in the SuperFoldAE model. Inspired by [13], this architecture incorporates a classification loss at the latent layer, enabling the model to optimize for both classification and reconstruction simultaneously. As a result, SuperFoldAE outperformed both DeepSF and PSI-BLAST, highlighting its effectiveness in protein fold classification. Our analysis utilizing the SCOP 1.75 dataset underscores the potential of AE-based representations in improving classification accuracy and reliability. The insights gained from this work contribute to our understanding of protein structures and pave the way for further exploration in representation-based deep learning.

For future research, we plan to extend our approach by applying it to independent datasets, allowing us to evaluate the robustness and generalizability of our model. Additionally, we aim to broaden the classification scope beyond protein folds to include family and superfamily classifications, providing a more comprehensive understanding of protein structure-function relationships. Furthermore, we intend to explore alternative Autoencoder architectures, such as graph-based Autoencoders [10], to capture structural information more effectively and improve the overall performance of protein classification tasks. This ongoing research aims to advance the field of computational biology and deepen our insights into the functional implications of protein structures.

Acknowledgments. The authors acknowledge the University of Maryland supercomputing resources (http://hpcc.umd.edu) made available for conducting the research reported in this paper.

References

1. Ahmed, S., et al.: Deep learning modelling techniques: current progress, applications, advantages, and challenges. Artif. Intell. Rev. **56** (2023). https://doi.org/10.1007/s10462-023-10466-8
2. Alam, F.F., Rahman, T., Shehu, A.: Learning reduced latent representations of protein structure data. In: Proceedings of the 10th ACM International Conference on Bioinformatics, Computational Biology and Health Informatics, pp. 592–597. BCB 2019, Association for Computing Machinery, New York, NY, USA (2019). https://doi.org/10.1145/3307339.3343866
3. Alam, F.F., Rahman, T., Shehu, A.: Evaluating autoencoder-based featurization and supervised learning for protein decoy selection. Molecules **25**(5) (2020). https://doi.org/10.3390/molecules25051146, https://www.mdpi.com/1420-3049/25/5/1146
4. Alam, F.F., Shehu, A.: Generating physically-realistic tertiary protein structures with deep latent variable models learning over experimentally-available structures. In: 2021 IEEE International Conference on Bioinformatics and Biomedicine (BIBM), pp. 2463–2470 (2021https://doi.org/10.1109/BIBM52615.2021.9669584
5. Alam, F.F., Shehu, A.: Data size and quality matter: generating physically-realistic distance maps of protein tertiary structures. Biomolecules **12**(7) (2022). https://doi.org/10.3390/biom12070908, https://www.mdpi.com/2218-273X/12/7/908
6. Berman, H., Henrick, K., Nakamura, H.: Announcing the worldwide protein data bank. Nat. Struct. Mol. Biol. **10**(12), 980–980 (2003)
7. Brown, W.M., Martin, S., Pollock, S.N., Coutsias, E.A., Watson, J.P.: Algorithmic dimensionality reduction for molecular structure analysis. J. Chem. Phys. **129**(6), 064118 (2008)
8. Chen, W., Tan, A.R., Ferguson, A.L.: Collective variable discovery and enhanced sampling using autoencoders: innovations in network architecture and error function design. J. Chem. Phys. **149**(7), 072312 (2018)
9. Cheng, J., Tegge, A.N., Baldi, P.: Machine learning methods for protein structure prediction. IEEE Rev. Biomed. Eng. **1**, 41–49 (2008). https://doi.org/10.1109/RBME.2008.2008239
10. Du, Y., et al.: Deep latent-variable models for controllable molecule generation. In: 2021 IEEE International Conference on Bioinformatics and Biomedicine (BIBM), pp. 372–375 (2021). https://doi.org/10.1109/BIBM52615.2021.9669692
11. Henikoff, S., Henikoff, J.G., Alford, W.J., Pietrokovski, S.: Automated construction and graphical presentation of protein blocks from unaligned sequences. Gene **163**(2), GC17–GC26 (1995)
12. Hou, J., Adhikari, B., Cheng, J.: DeepSF: deep convolutional neural network for mapping protein sequences to folds. Bioinformatics **34**(8), 1295–1303 (2017). https://doi.org/10.1093/bioinformatics/btx780, https://doi.org/10.1093/bioinformatics/btx780
13. Le, L., Patterson, A., White, M.: Supervised autoencoders: Improving generalization performance with unsupervised regularizers. In: Bengio, S., Wallach, H., Larochelle, H., Grauman, K., Cesa-Bianchi, N., Garnett, R. (eds.) Advances in Neural Information Processing Systems, vol. 31. Curran Associates, Inc. (2018)
14. Lemke, T., Peter, C.: Encodermap: dimensionality reduction and generation of molecule conformations. J. Chem. Theory Comput. **15**(2), 1209–1215 (2019)
15. Lyons, J., Paliwal, K.K., Dehzangi, A., Heffernan, R., Tsunoda, T., Sharma, A.: Protein fold recognition using hmm-hmm alignment and dynamic programming. J. Theor. Biol. **393**, 67–74 (2016)

16. Murzin, A.G., Brenner, S.E., Hubbard, T., Chothia, C.: SCOP: a structural classification of proteins database for the investigation of sequences and structures. J. Mol. Biol. **247**(4), 536–540 (1995)
17. Paszke, A., et al.: Pytorch: An imperative style, high-performance deep learning library. In: Advances in Neural Information Processing Systems, vol. 32, pp. 8024–8035. Curran Associates, Inc. (2019). http://papers.neurips.cc/paper/9015-pytorch-an-imperative-style-high-performance-deep-learning-library.pdf
18. Pedregosa, F., et al.: Scikit-learn: machine learning in Python. J. Mach. Learn. Res. **12**, 2825–2830 (2011)
19. Rahman, T., Alam, F.F., Shehu, A.: Equivariant encoding based GVAE (EqEn-GVAE) for protein tertiary structure generation. In: 2022 IEEE International Conference on Bioinformatics and Biomedicine (BIBM), pp. 3470–3477 (2022). https://doi.org/10.1109/BIBM55620.2022.9995502
20. Wei, L., Liao*, M., Gao, X., Zou, Q.: Enhanced protein fold prediction method through a novel feature extraction technique. IEEE Trans. NanoBiosci. **14**(6), 649–659 (2015). https://doi.org/10.1109/TNB.2015.2450233
21. Xia, J., Peng, Z., Qi, D., Mu, H., Yang, J.: An ensemble approach to protein fold classification by integration of template-based assignment and support vector machine classifier. Bioinformatics **33**(6), 863–870 (2017)

Enhancing Drug Discovery via Physics-Guided Deep Generative Models

Dikshant Sagar[1,2], Ari Jasko[2], and Negin Forouzesh[2(✉)]

[1] Donald Bren School of Information and Computer Sciences,
University of California, Irvine, Irvine CA 92617, USA
dikshans@uci.edu

[2] Department of Computer Science, California State University, Los Angeles,
Los Angeles CA 90032, USA
{ajasko,neginf}@calstatela.edu

Abstract. In the pursuit of structure-based drug discovery, the goal is to find small molecules capable of binding to a particular target protein and altering its function. Recently, deep learning (DL) has emerged as a promising approach for crafting drug-like molecules. It excels in creating compounds possessing precise biochemical characteristics while being influenced by structural features. Yet, their typical shortfall lies in the neglect of a critical element: the intrinsic physics that governs the structure and binding of molecules within real-world contexts. In this study, we explore and build on deep generative models informed by physics principles for drug discovery. These models not only consider the binding site but also incorporate physics-derived features that describe the interaction mechanism between a receptor and a ligand. We tested the proposed models by generating corresponding drug molecule candidates for a variety of protein-ligand complexes from the PDBBind dataset. On average, more than 75% of the structures generated by our hybrid model were stronger binders than the original experimental reference ligands to the protein. In addition, they had higher values of ΔG_{bind} (binding affinity) than molecules generated by the baseline methods by an average margin of 1.39 kcal/mol. Moreover, drug-like attributes of the generated molecules are evaluated in accordance with the Lipinski rules. To extend the analysis, their synthesizability is evaluated using ASKCOS, elevating the evaluation to a more comprehensive level. This revealed that the hybrid models notably excel in generating synthesizable molecules, with scores suggesting a higher likelihood of successful synthesis. Adherence to the Lipinski Rule of Five was also high, with compliance of 98.9%, suggesting favorable drug-like properties and a reduced risk of development failure due to poor bioavailability. This approach outperforms previous works, indicating significant improvements in drug discovery by enhancing both binding affinity and synthesizability.

Keywords: Drug discovery · Deep learning · Generative neural networks · physics-based simulation

1 Introduction

Drug discovery is vital for improving healthcare outcomes, but has traditionally been expensive (1.8 Billion) [38], slow (10 to 15 years) [6], and resource-intensive, with failure rates of over 90% [28,40]. Finding drug candidates typically takes 3 to 5 years. However, with the advancement of new computational techniques, this timeframe could be reduced to just a few weeks [23]. At the core of drug action is the interaction between a protein and its ligand-a small drug molecule. This crucial event involves molecular recognition and dynamic interplay within the protein's active site or binding pocket [26].

The search space for possible molecular structures is enormous and complex. It is estimated that there are 10^{60} drug-like structures possible [35]. It can be narrowed down by validating candidate molecules based on their chemical constraints, such as bond orders, molecular conformation, and valences. This search space shrinks significantly when the objective is to find a suitable molecule that precisely fits a designated binding pocket for goals such as receptor inhibition for disease therapy, and targeted drug delivery mechanisms. The discovery of a new molecular structure is divided into two main steps: (1) identifying promising compounds within a defined chemical space and (2) verifying their ability to bind to the target site as predicted. The first step can take anywhere from three to five years, rendering it a highly time-intensive process [3]. This deficiency highlights the demand for computational systems capable of intelligently navigating this restricted chemical search space and conducting virtual screenings of compounds for their potential to bind successfully. Implementing such systems could lead to considerable reductions in cost and expedite the drug development process.

Artificial Intelligence (AI) and Deep Learning (DL) have been on track to revolutionize computer-aided drug design by enhancing the efficiency and accuracy of the drug discovery process. These technologies can process and analyze vast datasets far beyond human capability, identifying patterns and insights that can lead to the discovery of new drug candidates. Some recent DL algorithms have been designed to predict molecular behavior, optimize drug properties, and simulate how drugs interact with biological targets, all with never seen before precision [16,20,22,32,34,36]. This ability to quickly generate and evaluate potential drug molecules can significantly reduce the time and cost associated with traditional drug development methods. Moreover, AI-driven models continuously learn and improve from new data, promising increasingly effective drug design strategies over time.

Although these recent developments marked a considerable advancement in the creation of new drug candidates, they did not account for essential physical attributes of the binding process. Specifically, they overlooked the protein-ligand ΔG_{bind}, encompassing both the enthalpic aspects (such as polar, non-polar, and Van der Waals energies) and the entropic components [48]. While structural information, such as bond connectivity and atom arrangements, forms the basis for molecular representations, they do not capture chemical systems' intricate and dynamic nature. A previous work [36], validated the effectiveness of integrating physics-based features into the molecular generation pipeline, which enhanced the binding affinity of the predicted ligands. In this research, we advance the prior achievements of [36] by first augmenting the design of the generative model. This enhancement involves integrating a discriminator component into the conditional variational autoencoder, thus evolving it into a sophisticated conditional variational generative adversarial network (CVAE-GAN) informed and directed by physics-based attributes. Furthermore, we conduct a more comprehensive analysis of the generated molecules' quality by evaluating their binding affinity for a much larger number of binding pockets from the PDBBind dataset. This evaluation is also expanded to include assessments of drug-likeness and synthesizability, offering a deeper insight into the potential practical applications of these molecules in therapeutics.

2 Related Work

Computer-Aided Drug Design (CADD) emerged to address Structure-Based Drug Discovery (SBDD) limitations, revolutionizing pharmaceutical discovery. This approach introduced innovative tools and methodologies that significantly accelerated drug development while reducing costs and risks [24]. [33] introduced the integration of Deep Learning (DL) into structure-based drug discovery, demonstrating its potential using convolutional neural networks (CNNs) [18] for pose-scoring functions. Subsequently, DL techniques have been applied to various tasks, from optimizing molecular poses [33] to predicting binding affinities [12], enhancing molecular docking strategies. However, these methods primarily extend to screening existing structures. Initial works [7,13,37] generated new molecular structures using DL with SMILES syntax [47] and graph-based representations [10,17,30,39], but struggled to fully capture the 3D molecular structure.

3D Molecular Representation. Traditional handling of molecular data representations in a 2D space can be counterintuitive, as it does not fully reflect the reality where molecular bonds have the ability to rotate, leading to diverse conformations of the molecule [43]. These various conformations play a crucial role in influencing inter-molecular interactions, such as the binding of a molecule to a receptor. To address these limitations, a 3D representation of molecules was developed through the use of atomic density grids [41]. Each voxel represents a distinct point in space in this system, meaning their identification relies on a coordinate framework. Moreover, they possess permutation invariance, which reduces the computational load required for comparisons. This makes them more efficient for analytical purposes. This approach better captures molecules' spatial configurations and complexities, accurately depicting their behavior and interactions.

Receptor Conditioned Molecule Generation. Previous studies [34] employed these 3D density grids for molecular representation and training a CVAE [42] with a conditional input protein receptor and ligand pairs in order to find novel structures. This approach allows for predicting new possible drug molecule structures that specifically bind to a particular binding pocket that is fed to the model at inference. The CVAE hybrid model [36] built upon this work to show how fusing physics-based information about the binding process of the protein-ligand pairs to the conditional input can improve the learning capabilities of the network which in turn generates better binders.

3 Materials and Methods

3.1 Dataset

For the purposes of this study, following the CVAE hybrid model [36], the primary dataset utilized is referred to as the PDBBind dataset [46]. This dataset is widely recognized within the scientific community for its comprehensive collection of known protein-ligand pairs, offering a rich resource for studies focused on drug discovery and molecular docking. In this work, we use a subset of the PDBBind-v19 known as the refined set, which has undergone additional optimization outlined in [45] to improve its quality and reliability for research purposes. Among the original 3,562 receptor-ligand complexes, 2,728 pairs had all the required features and experimental values available. The dataset was then divided into training and testing subsets in a random fashion, adhering to an 80:20 split. The purpose behind segregating a testing set was to eliminate the risk of overfitting and to assess the efficacy of the model's training by analyzing its performance through loss metrics. Ultimately, proteins from the test set were further used to predict drug molecule candidates. These candidates were subsequently assessed based on the evaluation metrics outlined in Sect. 4.2, providing a comprehensive evaluation of their quality and viability.

3.2 Physics-Based Features

Implicit solvent modeling is one of the most popular computational methods that consider the solvent (usually water) as one continuum component. Within this framework, the calculation of binding free energy (ΔG_{bind}) could be conducted more efficiently compared to other computational models, *e.g.*, explicit solvents. Poisson-Boltzmann (PB) and generalized Born (GB) models are the two main classes of implicit solvent models that have been used widely in static and dynamic simulations of protein-ligand interactions [31]. In this work, GBNSR6 [8,9] and PBSA [21] in AmberTools20 [5] are used for fast yet accurate calculation of ΔG_{bind} (see Table 1). By integrating implicit solvents into the DL model, it is more likely to generate feasible and strong binders.

Table 1. Physics-based features calculated for complex, protein, and ligand structures using MM/PB(GB)SA tool. *Entropy* is calculated as the difference between the experimental ΔG_{bind} and computational Enthalpy values. See [4] for details.

Parameter	Description	Method	Count
1-4-EELEC	1-4 Electrostatic energy	GB	3
VDWAALS	Van der Waals energy	PB	3
EELEC	Electrostatic energy	GB&PB	6
ESURF	Non-polar solvation energy	GB	3
EGB	Polar solvation energy	GB	3
ECAVITY	Non-polar solvation free energy	PB	3
EPB	Reaction field energy	PB	3
Etot	Computational calculated $\Delta\Delta G$	GB&PB	6
Enthalpy	Total energy of a system	GB	1
Entropy	Entropy	E^*	1
ΔG_{bind}	Binding free energy	GB	1

3.3 Atomic Vector Representation

To facilitate the training of the deep generative neural network, molecular data has to undergo transformation into a vector format. This process involves representing each atom as an individual vector, resulting in each molecule being depicted as a collection of atom-type vectors. We follow the same atom typing scheme as described in the CVAE and CVAE hybrid models [34,36], where atom types are assigned using a set of N_p atomic property functions **p** and value ranges for those properties v as done in [34,36]. The atomic properties used here were element (different value ranges for ligands and receptors), aromaticity, H bond donor and acceptor status, and formal charge. For every atom a, a one-hot encoded vector **p** is created for each property, and then N_p vectors are concatenated to create a final atom type vector $t \in \mathbb{R}^{N_t}$. Hence, we get a 1×18 sized vector per atom.

3.4 Molecule Density Grid Representation

Once a molecule has been atom-typed, choosing a representation that captures its 3D spatial features becomes crucial. Therefore, we utilized a molecular gridding library called libmolgrid [44] that creates a density grid representation of molecules where atoms are represented as continuous densities with truncated Gaussian shapes. Libmolgrid defines the density value of an atom at a grid point by a kernel function $f : \mathbb{R} \times \mathbb{R} \to \mathbb{R}$ that takes as input the distance d between the atom coordinate and the grid point and the atomic radius r:

$$f(d,r) = \begin{cases} e^{-2(\frac{d}{r})^2}, & d \leq 1.5\ r \\ 0, & d > 1.5\ r \end{cases} \quad (1)$$

r was fixed to 1.0 Å for all atoms, and the dimension of the cubic grid to 23.5 Å with 0.5 Å resolution to maintain consistency with [34,36], which results in spatial dimensions of $N_X = N_Y = N_Z = 48$. Also, N is the total number of atoms. In order to conserve computational resources, only those atoms that fall within the spatial boundaries of the grid are included in the representation.

3.5 Atom Fitting and Bond Inference

Given that our generative models are developed and trained using data in the format of molecular density grids, the model's predictive output similarly manifests as density grid representations. Now the problem remains of converting a reference density grid G_{ref} back into a discrete 3D molecular structure, which does not have an analytical solution [34] and is therefore solved with the following optimization problem:

$$\mathbf{T}^*, \mathbf{C}^* = \underset{T,C}{\operatorname{argmin}} \|\mathbf{G_{ref}} - g(\mathbf{T}, \mathbf{C})\|^2 \tag{2}$$

where g is the function to convert a molecule's atom type vector T and atomic coordinate vector C into density grid G. The initial locations of atoms can be found by selecting the grid points with the largest density values. By using libmolgrid, we can compute the grid representation of an atomic structure and backpropagate a gradient from grid values to atomic coordinates. We use the algorithm defined in [34] that combines iterative atom detection with gradient descent to find the best set of atoms that fit that reference density grid. After identifying the atoms and their coordinates, the remaining step involves establishing bonds among the atoms to create valid molecules. This process is facilitated by a bond inference algorithm that uses a sequence of inference rules that add bond information and hydrogens while trying to satisfy the constraints defined by the atom types, a customized bond perception routine developed within the OpenBabel framework [29] (Fig. 1).

3.6 Physics-Guided Deep Generative Hybrid Model

Previous work [36] introduced and started a family of sophisticated deep generative models that leveraged physics-informed guidance and improved the new molecule generation process. The model introduced was built upon a Conditional Variational Autoencoder (CVAE) framework. During its training phase, it processes the molecular density grid representations of both the conditional receptor protein's binding pocket and the reference ligand. This is done alongside integrating the physics-based characteristics of the interacting pair, enhancing the model's predictive accuracy and relevance in simulating molecular interactions, which, in turn, better guides the generation process. The objective was to learn a sample from a distribution $p(lig|rec, feat)$ where lig, rec, and $feat$ are the ligand density grid, receptor density grid, and physics-based features, respectively. The latent sample z was drawn from a standard normal distribution under the assumption that the binding interactions might follow it as a

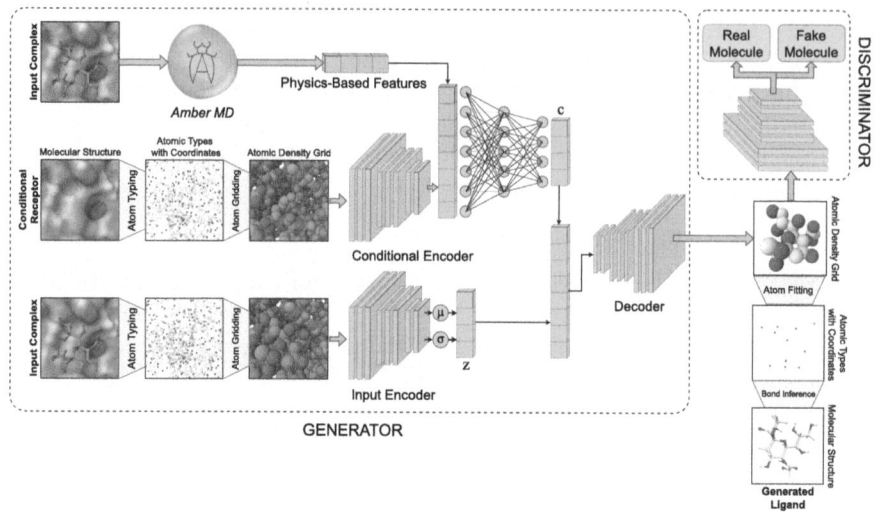

Fig. 1. Our physics-guided deep generative model pipeline overview. First, the docked protein and ligand complex are transformed into atom-type vectors, which are then converted into atomic density grids. Following this, the encoder branches of our physics-informed CVAEGAN model process the input complex alongside the density grids of the protein receptor and incorporate the physics-derived features. The input the encoder produces a probabilistic latent vector sampled from $z \sim N(\mu, \sigma)$, and the conditional encoder gives an encoded vector c, which is then concatenated to z and fed into the decoder to produce an output-generated ligand density grid. This generated molecular grid density is then fed to our discriminator subnetwork to classify it as real or fake. The molecular density grid is then finally converted to a 3D molecular structure by atom fitting and bond inference algorithms.

prior. In the generative process, they first drew a sample $z \sim p(z)$ and then generated $lig_{gen} \sim p_\theta(lig|z, c)$, where p_θ is the same decoder neural network and c is the encoding from the conditional encoder (Fig. 2). To form the basis of this incremental work, we have our input complex encoder that takes the molecular density grid representations calculated using libmolgrid [44] as inputs. The input encoder transforms the inputs, specifically the receptor rec and ligand lig, into a defined set of means μ and standard deviations σ similar to the CVAE and CVAE hybrid models [34,36]. These parameters delineate the latent variables from which a latent vector z is sampled. Simultaneously, the conditional encoder operates by mapping the identical receptor protein rec alongside the physics-based features $feat$ into a conditional encoding vector c. This process encapsulates the contextual information provided by both the receptor and its associated physical characteristics into a unified representation for guiding the generative process. Following this, the concatenated vector of z and c is fed into the decoder network. The decoder then processes this concatenated vector, ultimately decoding it into a generated molecular density grid lig_{gen} representing the generated ligand molecule.

Now, to enhance the quality of the generated molecular density grid representations, we integrated a sub-neural network functioning as a discriminator to the whole pipeline. This discriminator network is tasked with distinguishing between real and fake density grids generated by the model throughout the training process. Concurrently, the CVAE assumes the role of a generator network. We adopt an adversarial training methodology akin to that normally utilized in training GANs [14], refining the entire network to produce more accurate and realistic molecular structures. Finally, lig_{gen} is passed through the discriminator network to receive a label of a fake or a real molecule, which in turn forces the generator network *(i.e., the CVAE)* to improve its performance by updating its weights and produce more realistic molecular density grids, thus improving the ultimate outcome. Hence, this network takes the form of a conditional variation autoencoder generative adversarial neural network (CVAEGAN) [2].

4 Experimental Setup

4.1 Training and Optimization

To train the CVAE or the generator in this pipeline we follow the previous work [36] where due to the difficult nature of estimating the naive maximum likelihood to compute the latent posterior probability $p_\theta(z|rec, lig)$, we learn an approximate input encoder model $q_\phi(lig|z, c)$ of the posterior distribution which can be trained by the following two objectives:

$$L_{recon} = -log\, p_\theta(lig|z,c) \propto \frac{1}{2}||lig - lig_{gen}||^2 \qquad (3)$$

$$L_{KL} = D_{KL}(q_\phi(z|lig,c)||p(z)) \qquad (4)$$

L_{recon} is the reconstruction loss term which maximizes the probability that the latent samples from the approximate posterior distribution $z \sim q_\phi(z|rec, lig, feat)$ are decoded as close to the original ligand density lig that was provided during the forward pass. L_{KL} is the Kullback-Liebler (KL) divergence loss that forces the learned latent space probability distribution to be as close as possible to a standard normal distribution, *i.e.*, $p(z) = N(0, 1)$. With the joint optimization of both these terms, we are able to learn a latent space that follows a normal distribution, and we end up training a decoder that can decode these latent vectors sampled from a normal distribution into realistic ligand densities. Similar to the CVAE and CVAE hybrid models [34,36], we also include the loss term called Steric Loss that minimizes steric clash in terms of the overlap between the generated molecular density and the receptor pocket density. The loss value is calculated by first summing across the grid channels, then multiplying the receptor and ligand density at each point:

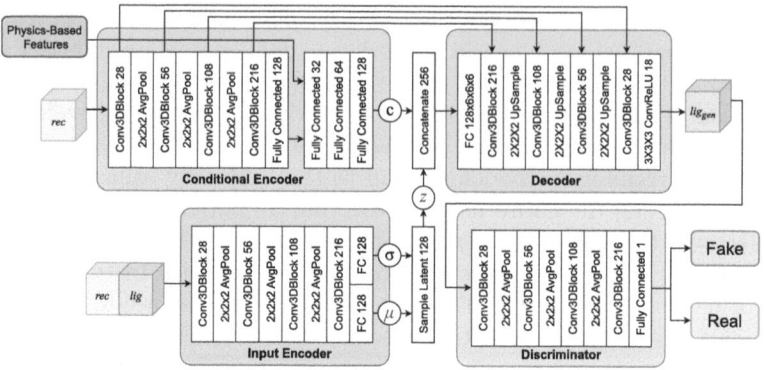

Fig. 2. Our physics-guided deep generative hybrid model's internal architecture.

$$L_{steric} = \left\langle \sum_i^{N_T} rec_i, \sum_i^{N_T} lig_{gen,i} \right\rangle \tag{5}$$

Training the new discriminator subnetwork, while utilizing the CVAE output as a generator, simplifies to a scenario where we aim to minimize a minmax loss. In this setup, the generator's objective is to synthesize molecules that are convincing enough for the discriminator to classify as real, implying they originate from an authentic distribution rather than being artificially generated. However, this training approach, akin to that used in GANs [14], suffers from issues such as mode collapse and vanishing gradients. To address these challenges, we employ the Wasserstein GAN loss approach [1]. Unlike traditional GANs, where the discriminator outputs probabilities, the Wasserstein approach assigns a clipped score to both real and generated molecules. Consequently, our objective shifts to minimizing the difference between these two scores, enhancing the stability and reliability of the training process. Therefore the discriminator loss term becomes:

$$L_{Disc} = D(x) - D(G(z|c)) \tag{6}$$

Hence, the final loss objective for the complete model becomes:

$$L = \lambda_{recon}L_{recon} + \lambda_{KL}L_{KL} + \lambda_{steric}L_{steric} + \lambda_{disc}L_{disc} \tag{7}$$

The loss weights were kept consistent with [34,36] at $\lambda_{recon} = 4.0$, $\lambda_{KL} = 0.1$, $\lambda_{steric} = 1.0$ and $\lambda_{disc} = 1.0$, with the KL divergence loss weight increased to 1.6 after 20,000 iterations. The model was fine-tuned using the RMSProp optimizer with gradient clipping with a learning rate of 10^{-7} for 100,000 iterations and a batch size of 4 using a computation cluster node with an NVIDIA A30 GPU.

4.2 Evaluation Metrics

In order to assess the quality of ligands generated by different approaches, we have adopted and applied a range of evaluation metrics. Similar to [34,36] as a base evaluation, we employed the metric known as ΔG_{bind} calculated using the GNINA package [25], which represents the binding affinity value between the receptor and ligand. A negative ΔG_{bind} value indicates a favorable binding interaction, suggesting a stronger affinity between the ligand and the receptor [11]. In our evaluation of drug candidates produced by our deep generative model for drug discovery, a critical factor we also consider is the synthesizability of these molecules. Synthesizability refers to the ease with which a molecule can be synthesized in a laboratory. This significantly affects a candidate's practicality, cost-effectiveness, and development timeline. To assess this quantitatively, we use ASKCOS [27], an advanced organic synthesis planning tool powered by a neural network trained on the comprehensive Reaxys dataset [15]. ASKCOS evaluates synthesizability through a heuristic score. A higher heuristic score indicates that a molecule is easier to synthesize, reflecting the neural network's learned patterns from extensive chemical reaction data. These insights are crucial for identifying the most viable candidates for efficient and cost-effective drug development. Within drug discovery, the main goal is to identify new drug molecules and evaluate their potential as effective treatments. This evaluation is based on their "drug-likeness" or "drugability" - key attributes that determine their suitability for therapeutic use. Therefore, we also try to assess these properties using a rule-based metric called Lipinski's Rule of Five [19], which is a preliminary screening tool in drug discovery that helps identify molecules that are likely to be orally bioavailable. However, while useful, these rules are not definitive; exceptions can still lead to successful drugs [49]. Lipinski's Rule of Five states that a compound is more likely to be absorbed if: (1) hydrogen bond donors (HBD) < 5, (2) hydrogen bond acceptors (HBA) < 10, (3) molecular mass $(m) < 500$ daltons, (4) octanol-water partition coefficient (log P) < 5.

5 Results

We compared our proposed model with two baselines: the CVAE model [34] and the CVAE hybrid model [36]. Using each model, we generated 90 unique molecules for every binding pocket in the testing set (546 proteins). We then selected the top 5 molecules per pocket based on binding affinities and assessed their structures using metrics for binding affinity, synthesizability, and drug-likeness defined in Sec.4.2. Table 2 presents averaged values for each metric, while Fig. 3 shows the distribution of these metrics and atom-type frequency analysis across models. We see that in terms of ΔG_{bind} (binding affinity), we surpass the previous baselines, achieving an average ΔG_{bind} of -10.70 kcal/mol, whereas the CVAE [36] and CVAE hybrid model [34] achieved -9.79 kcal/mol and -8.91 kcal/mol, respectively. This outcome aligns with our expectations, given that the neural network is conditioned on the reference ligand and incorporates its physics-based features. Furthermore, as hypothesized, the discriminator sub-network improves the quality of the generated molecular density grids, thus enhancing the output's sharpness. Such conditioning enables the network to specifically generate molecules with enhanced binding affinity. Now, in

Table 2. Comparing the average metric values for molecules generated by the models for each binding pocket in the test set.

Model	ΔG_{bind} (↓)	ASKCOS Sc.(↑)	m	$Log P$	HBA	HBD
CVAE [34]	−8.91	−2.33e4	321.04	1.15	5.05	3.75
CVAE Hybrid [36]	−9.79	**−1.75e5**	310.07	−0.62	6.73	4.8
CVAEGAN Hybrid	**−10.70**	−2.17e4	358.47	−0.35	7.42	4.91

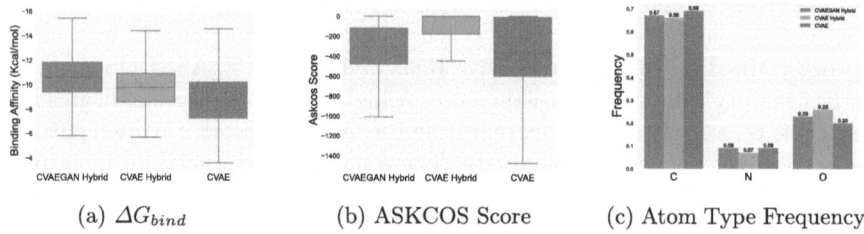

(a) ΔG_{bind} (b) ASKCOS Score   (c) Atom Type Frequency

Fig. 3. Box plots illustrating and comparing the distribution of (a) ΔG_{bind} (kcal/mol), (b) ASKCOS scores, and (c) Relative atom type frequencies for all molecules generated by the three models for the binding pockets in the test set.

terms of synthesizability, we achieve comparable ASKCOS scores compared to the baselines, achieving the second-best average ASKCOS score of −1.75e5. It's important to recognize that ASKCOS calculates synthesizability scores based on known precursors and compounds. This methodology can yield unusual and unrepresentative scores when dealing with the generation of novel compounds that have never been previously encountered or recorded by a system like ASKCOS. We believe this could be the reason for the unusually high negative scores observed across all three models. Lastly, regarding drug-likeness, the metrics are similar across all models, particularly considering that neither the baselines nor our method explicitly impose or condition adherence to Lipinski's rules during training. We believe that the CVAE model [34] achieving slightly better Lipinski values is attributable to the extensive size of their dataset. This includes nearly 22.5 million protein-ligand pairs. This vast collection likely encompasses a more diverse and druglike set of structures, implicitly conditioning the model to generate molecules that more closely adhere to Lipinski's rules. In contrast, the CVAE hybrid model [36] and our model were trained on approximately 2,100 protein-ligand pairs from the PDBBind dataset, constrained by the availability of physics-based features specific to these entries. This difference in dataset scope and content significantly influences the training outcomes and the drug-likeness of the generated molecules.

In Fig. 4, we visualize the top-5 generated molecules by the three models inside a single binding pocket of the protein 1igb [beta-d-glucan glucohydrolase isoenzyme exo1] from the PDBBind test set. We observed that docked molecules predicted by our CVAEGAN hybrid model have more feasible conformation and

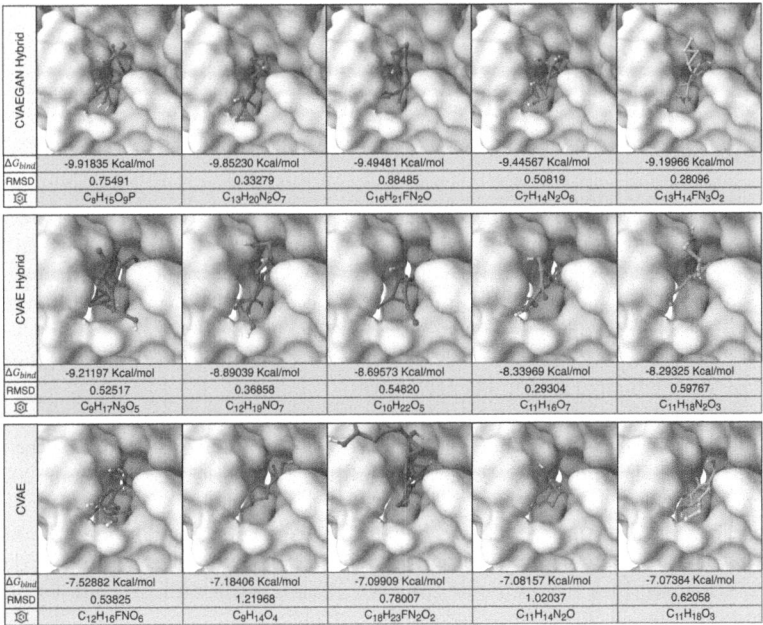

Fig. 4. Visualisation of the Top-5 generated ligands inside the receptor pocket by the baselines - LiGAN [33] and CVAE Hybrid [36] in comparison to our physics-guided deep generative hybrid model - CVAEGAN Hybrid for the PDBBind protein 1igb [beta-d-glucan glucohydrolase isoenzyme exo1] with their decreasing binding affinities (left to right) and their RMSD values.

orientation inside the protein binding pocket. The corresponding better ΔG_{bind} values also confirm that they are better binders all across.

6 Conclusion and Future Work

In this work, we demonstrated that incorporating physics-based data into deep generative models enhances their ability to predict superior molecular structures for receptor proteins. This approach not only demonstrates great potential but also promises to transform the field. By merging DL techniques with core physical principles, we have successfully advanced traditional drug discovery methods. Our hybrid physics-based CVAEGAN model generated realistic structures with higher ΔG_{bind} compared to computational baselines and known reference ligands. The model also successfully predicted synthesizable and drug-like molecules. However, we identified a key limitation: the lack of explicit conditioning for drug-likeness. Future studies could enhance the model by integrating drug-likeness and synthesizability constraints, improving its practical applications in drug development.

Acknowledgment. This research has been funded by NIH R16GM146633 and NSF 2216858 grants awarded to N.F.

References

1. Arjovsky, M., Chintala, S., Bottou, L.: Wasserstein generative adversarial networks. In: International Conference on Machine Learning, pp. 214–223. PMLR (2017)
2. Bao, J., Chen, D., Wen, F., Li, H., Hua, G.: CVAE-GAN: fine-grained image generation through asymmetric training. In: Proceedings of the IEEE International Conference on Computer Vision, pp. 2745–2754 (2017)
3. Berdigaliyev, N., Aljofan, M.: An overview of drug discovery and development. Future Med. Chem. **12**(10), 939–947 (2020)
4. Cain, S., Risheh, A., Forouzesh, N.: A physics-guided neural network for predicting protein–ligand binding free energy: from host–guest systems to the PDB-bind database. Biomolecules **12**(7) (2022). https://doi.org/10.3390/biom12070919, https://www.mdpi.com/2218-273X/12/7/919
5. Case, D.A., et al.: Amber 2021. University of California, San Francisco (2021)
6. Derep, M.: What's the average time to bring a drug to market in 2022? https://lifesciences.n-side.com/blog/what-is-the-average-time-to-bring-a-drug-to-market-in-2022. Accessed 10 May 2024
7. Ertl, P., Lewis, R., Martin, E., Polyakov, V.: In silico generation of novel, drug-like chemical matter using the LSTM neural network. arXiv preprint arXiv:1712.07449 (2017)
8. Forouzesh, N., Izadi, S., Onufriev, A.V.: Grid-based surface generalized born model for calculation of electrostatic binding free energies. J. Chem. Inf. Model. **57**(10), 2505–2513 (2017)
9. Forouzesh, N., Mukhopadhyay, A., Watson, L.T., Onufriev, A.V.: Multidimensional global optimization and robustness analysis in the context of protein-ligand binding. J. Chem. Theory Comput. **16**, 4669–4684 (2020)
10. Gilmer, J., Schoenholz, S.S., Riley, P.F., Vinyals, O., Dahl, G.E.: Neural message passing for quantum chemistry. In: International Conference on Machine Learning, pp. 1263–1272. PMLR (2017)
11. Gilson, M.K., Zhou, H.X.: Calculation of protein-ligand binding affinities. Annu. Rev. Biophys. Biomol. Struct. **36**, 21–42 (2007)
12. Gomes, J., Ramsundar, B., Feinberg, E.N., Pande, V.S.: Atomic convolutional networks for predicting protein-ligand binding affinity. arXiv preprint arXiv:1703.10603 (2017)
13. Gómez-Bombarelli, R., et al.: Automatic chemical design using a data-driven continuous representation of molecules. ACS Cent. Sci. **4**(2), 268–276 (2018)
14. Goodfellow, I., et al.: Generative adversarial nets. In: Advances in Neural Information Processing Systems, vol. 27 (2014)
15. Goodman, J.: Computer software review: Reaxys (2009)
16. Rao, V.S., Srinivas, K.: Modern drug discovery process: an in silico approach. J. Bioinfo. Seq. Anal. **2**(5), 89–94 (2011)
17. Sun, D., Gao, W., Hu, H., Zhou, S.: Why 90% of clinical drug development fails and how to improve it? Acta Pharm. Sinica B **12**(7), 3049–3062 (2022)
18. LeCun, Y., Bengio, Y., et al.: Convolutional networks for images, speech, and time series. In: The handbook of Brain Theory and Neural Networks, vol. 3361, no. 10 (1995)

19. Lipinski, C.A., Lombardo, F., Dominy, B.W., Feeney, P.J.: Experimental and computational approaches to estimate solubility and permeability in drug discovery and development settings. Adv. Drug Deliv. Rev. **23**(1–3), 3–25 (1997)
20. Liu, M., Luo, Y., Uchino, K., Maruhashi, K., Ji, S.: Generating 3D molecules for target protein binding. In: International Conference on Machine Learning, pp. 13912–13924. PMLR (2022)
21. Luo, R., David, L., Gilson, M.K.: Accelerated poisson-boltzmann calculations for static and dynamic systems. J. Comput. Chem. **23**(13), 1244–1253 (2002)
22. Luo, S., Guan, J., Ma, J., Peng, J.: A 3D generative model for structure-based drug design. Adv. Neural. Inf. Process. Syst. **34**, 6229–6239 (2021)
23. Mak, K.K., Wong, Y.H., Pichika, M.R.: Artificial intelligence in drug discovery and development. In: Drug Discovery and Evaluation: Safety and Pharmacokinetic Assays, pp. 1–38 (2023). https://doi.org/10.1007/978-3-031-35529-5_92
24. Marshall, G.R.: Computer-aided drug design. Annu. Rev. Pharmacol. Toxicol. **27**(1), 193–213 (1987)
25. McNutt, A.T., et al.: Gnina 1.0: molecular docking with deep learning. J. Chem. **13**(1), 1–20 (2021)
26. Mishra, N., Forouzesh, N.: Protein-ligand binding with applications in molecular docking. In: Algorithms and Methods in Structural Bioinformatics, pp. 1–16. Springer (2012). https://doi.org/10.1007/978-3-031-05914-8_1
27. MIT: ASKCOS - Automated system for knowledge-based continuous organic synthesis. https://askcos.mit.edu/. Accessed 15 April 2024
28. Mohs, R.C., Greig, N.H.: Drug discovery and development: role of basic biological research. Alzheimer's Dementia: Transl. Res. Clin. Inter. **3**(4), 651–657 (2017)
29. O'Boyle, N.: Banck m. james ca morley c. vandermeersch t. hutchison gr open babel: an open chemical toolbox. J. Cheminf **3**(1), 33 (2011)
30. Olivecrona, M., Blaschke, T., Engkvist, O., Chen, H.: Molecular de-novo design through deep reinforcement learning. J. Chem. **9**(1), 1–14 (2017)
31. Onufriev, A.: Implicit solvent models in molecular dynamics simulations: a brief overview. Ann. Rep. Comput. Chem. **4**, 125–137 (2008)
32. Peng, X., Luo, S., Guan, J., Xie, Q., Peng, J., Ma, J.: Pocket2mol: efficient molecular sampling based on 3D protein pockets. In: International Conference on Machine Learning, pp. 17644–17655. PMLR (2022)
33. Ragoza, M., Hochuli, J., Idrobo, E., Sunseri, J., Koes, D.R.: Protein-ligand scoring with convolutional neural networks. J. Chem. Inf. Model. **57**(4), 942–957 (2017)
34. Ragoza, M., Masuda, T., Koes, D.R.: Generating 3D molecules conditional on receptor binding sites with deep generative models. Chem. Sci. **13**(9), 2701–2713 (2022)
35. Reymond, J.L.: The chemical space project. Acc. Chem. Res. **48**(3), 722–730 (2015)
36. Sagar, D., Risheh, A., Sheikh, N., Forouzesh, N.: Physics-guided deep generative model for new ligand discovery. In: Proceedings of the 14th ACM International Conference on Bioinformatics, Computational Biology, and Health Informatics, pp. 1–9 (2023)
37. Segler, M.H., Kogej, T., Tyrchan, C., Waller, M.P.: Generating focused molecule libraries for drug discovery with recurrent neural networks. ACS Cent. Sci. **4**(1), 120–131 (2018)
38. da Silva, M.F., dos Santos, A.B.S.F., de Melo Batista, V., da Silva Rodrigues, É.E., de Araújo-Júnior, J.X., da Silva-Júnior, E.F.: New drug discovery and development. In: Dosage Forms, Formulation Developments and Regulations, pp. 3–65. Elsevier (2024)

39. Simonovsky, M., Komodakis, N.: GraphVAE: towards generation of small graphs using variational autoencoders. In: Kůrková, V., Manolopoulos, Y., Hammer, B., Iliadis, L., Maglogiannis, I. (eds.) ICANN 2018. LNCS, vol. 11139, pp. 412–422. Springer, Cham (2018). https://doi.org/10.1007/978-3-030-01418-6_41
40. SITNFlash: Modern drug discovery: why is the drug development pipeline full of expensive failures? (2020). https://sitn.hms.harvard.edu/flash/2020/modern-drug-discovery-why-is-the-drug-development-pipeline-full-of-expensive-failures/
41. Skalic, M., Sabbadin, D., Sattarov, B., Sciabola, S., De Fabritiis, G.: From target to drug: generative modeling for the multimodal structure-based ligand design. Mol. Pharm. **16**(10), 4282–4291 (2019)
42. Sohn, K., Lee, H., Yan, X.: Learning structured output representation using deep conditional generative models. In: Advances in Neural Information Processing Systems, vol. 28 (2015)
43. Stockwell, G.R., Thornton, J.M.: Conformational diversity of ligands bound to proteins. J. Mol. Biol. **356**(4), 928–944 (2006)
44. Sunseri, J., Koes, D.R.: Libmolgrid: graphics processing unit accelerated molecular gridding for deep learning applications. J. Chem. Inf. Model. **60**(3), 1079–1084 (2020)
45. Wang, B., Zhao, Z., Nguyen, D.D., Wei, G.W.: Feature functional theory-binding predictor (FFT-BP) for the blind prediction of binding free energies. Theoret. Chem. Acc. **136**, 1–22 (2017)
46. Wang, R., Fang, X., Lu, Y., Wang, S.: The PDBbind database: collection of binding affinities for protein- ligand complexes with known three-dimensional structures. J. Med. Chem. **47**(12), 2977–2980 (2004)
47. Weininger, D.: Smiles, a chemical language and information system. 1. introduction to methodology and encoding rules. J. Chem. Inf. Comput. Sci. **28**(1), 31–36 (1988)
48. Woo, H.J., Roux, B.: Calculation of absolute protein-ligand binding free energy from computer simulations. Proc. Natl. Acad. Sci. **102**(19), 6825–6830 (2005)
49. Zhang, M.Q., Wilkinson, B.: Drug discovery beyond the 'rule-of-five'. Curr. Opin. Biotechnol. **18**(6), 478–488 (2007)

Decoys Reveal Multiple Basins of Attraction for Cryo-Electron-Microscopy Flexible Fitting

Maytha Alshammari[1], Jing He[1], and Willy Wriggers[2](✉)

[1] Department of Computer Science, Old Dominion University, Norfolk, VA 23529, USA
[2] Department of Mechanical and Aerospace Engineering, Old Dominion University, Norfolk, VA 23529, USA
wriggers@biomachina.org

Abstract. This study explored the robustness and uniqueness of the flexible fitting of atomic structures against cryo-electron microscopy (cryo-EM) maps using elastic network motion models. The success of flexible fitting is based on the optimistic expectation of a single optimum fit that can be reached from a wide range of start conformations. We revisited this assumption for four AlphaFold models that deviated from corresponding medium-resolution cryo-EM maps but benefitted from flexible fitting. To test the dependence of the flexible fitting performance on the start structures, we systematically generated decoys using normal modes, offering a broader sampling of the conformational space compared to a single start structure. This strategy allowed exploration of the global properties of the cross-correlation (CC) scoring function landscape. Statistical analysis using multidimensional scaling revealed that the initial decoy ensembles collapsed into multiple basins of attraction in three of the four cases. The results demonstrate that a single start structure can be trapped in the local maxima of the CC during flexible fitting (spurious fits), but the decoys increase the likelihood of finding a correct fit. More precisely, there is a "winning" cluster of closely related structures that exhibit high template modeling (TM)-scores with the known true structures. Comparison of the CC and the TM-scores showed that the winning cluster can be identified by high CC values, further demonstrating the utility of cryo-EM maps as filters for screening candidate structures.

Keywords: Flexible Fitting · Cryo-EM · AlphaFold

1 Introduction

High-accuracy AlphaFold-predicted models [1] can provide detailed atomic predictions that complement lower-resolution cryo-electron microscopy (cryo-EM) maps, which cannot resolve individual atoms [2–4]. Recently, we have shown that AlphaFold models can benefit from flexible refinement [2–4], but atomic simulations may not provide sufficient reach when refined against deviating cryo-EM maps [5]. To reduce the dimensionality of the search and to allow larger-scale collective motions, we utilized normal mode analysis [6]. A coarse-grained elastic network model (ENM) was implemented to generate an exhaustive set of structure decoys from discretely sampled modes

[7]. ENMs use only the lowest-frequency normal modes, yielding a more manageable (lower-dimensional) representation of conformational changes while ignoring local (i.e., high-spatial-frequency) deformations. Using decoys as start structures, we observed a significant improvement in accuracy, with the best decoy exhibiting a TM-score of 0.68 compared to the original AlphaFold TM-score of 0.52 [7].

Encouraged by this success, we developed a new tool, *elforge.py*, in ModeHunter version 1.4 [5]. This tool optimizes a "continuously varying" decoy (a deformable model of the protein) and matches it with the cryo-EM map. This match is based on a search for the optimal mode elongation vector that maximizes the cross-correlation (CC) score between a simulated map generated from the deformed structures and the target map. In [5], we explored four optimization methods, including two local methods (Powell and Nelder-Mead) and two global methods (Dual Annealing and Differential Evolution). Moreover, we tested three different normal mode ranges (modes 1–9, 7–9, and 1–12), masked and unmasked (box-cropped) density maps, and two CC similarity measures (Pearson and inner product). Our aim in [5] was to find the best-performing conditions and parameter choices for the new *elforge.py* tool. Across four challenging AlphaFold systems, significant improvements in structure accuracy were observed when masked maps, modes 1–12, inner product, and Powell optimization were used, as evidenced by significant increases in TM-scores after flexible fitting [5]. The tests in [5] were based on specific start structures, those with the highest predicted local distance difference test (pLDDT) score used by AlphaFold, which we found to be within the reach of the global maximum of the CC-score used by flexible fitting.

In the present work, to reduce any biases by a single start structure, we combined the benefits of decoy generation [7] and ENM flexible fitting [5] to further explore the robustness and convergence properties. We modified the approach in [5] to start the refinement from multiple decoys (generated from the highest pLDDT-score AlphaFold models with the ModeHunter *eldecoy.py* tool [7]) to systematically produce a diverse ensemble of start conformations [7]. The new "multi-shot" approach to flexible fitting offers improved coverage of the search space, thereby enhancing the likelihood of achieving accurate model refinement. Using only local Powell optimization, it also avoids the inefficiencies and convergence problems of global optimization methods observed in [5].

To investigate the impact of starting with decoys, we first created, using *eldecoy.py*, an ensemble of 1,000 decoys from modes 7–9, with 10 sampling points for each degree of freedom and a maximum root mean square deviation (RMSD) scale of 5Å for each mode [7]. Then, we flexibly fitted each decoy into a cryo-EM density map using *elforge.py* [5]. The best parameters were observed in [5]. To visualize the structures before and after the fitting, we embedded the initial decoy ensembles and the fitting results on a 2D plane via multidimensional scaling (MDS), as implemented in the scikit-learn Python package [8]. Decoys after flexible fitting were clustered using the density-based spatial clustering of applications with noise (DBSCAN) algorithm, also from the scikit-learn Python package [9], to provide a measure of the precision of the fitting. Finally, we analyzed the TM-score and the CC-score as measures of accuracy to explore the predictive value of the CC in the screening of the resulting decoy refinements.

In the following section, we describe in detail mode generation using the ModeHunter tool *enmhunt.py*, AlphaFold decoy generation using *eldecoy.py*, flexible fitting using *elforge.py*, and clustering and evaluation of the fitted models. Then, we present the results of the multi-shot approach using the four test cases from [5]. Finally, we discuss the implications of our findings and the potential for integrating these methods to overcome the limitations of single model refinement.

2 Methods

We evaluated the performance of our multi-shot method using a test set that consisted of four protein chains (Sect. 2.1). AlphaFold decoys were generated using normal modes with ENM, employing *enmhunt.py* and *eldecoy.py* (Sect. 2.2), and were utilized for flexible fitting into cryo-EM for refinement (Sect. 2.3). The decoys and fitted structures were dimensionally reduced and clustered using MDS and DBSCAN (Sect. 2.4).

2.1 Data Preparation and Structure Prediction with AlphaFold

We used the four systems from our earlier study [5] to evaluate our multi-shot approach. Each test system consisted of an amino acid sequence, the known atomic structure, and a cryo-EM density map. In the following we provide only a brief overview of these four cases. For full structure details and molecular graphics see [5].

Lipid-Preserved Respiratory Supercomplex: The atomic structure 7DGQ chain 3 was retrieved from the Protein Data Bank (PDB) in June 2022. The associated electron density map, EMD 30673, with a resolution of 5Å, was acquired from the Electron Microscopy Data Bank (EMDB).

Flagellar L-Ring Protein: The atomic structure 7BGL chain A (CASP14 free modeling case T1047s1-D1) was downloaded from the PDB in March 2022. The resolution of the corresponding cryo-EM map (EMD 12183) was lowered to 5Å using our hybrid approach [4].

Cation Diffusion Facilitator YiiP: The atomic structure 7KZZ chain B was obtained from the repository of Terwilliger et al. [2] and downloaded in August 2021. A 5Å-resolution hybrid map was created from EMD 23093 [4].

***Sulfolobus Islandicus* Pilus**: The atomic structure 6NAV chain A was retrieved from the PDB in March 2023. The corresponding 4.1Å-resolution EMD 0397 map was obtained from the EMDB.

The models for the first three test systems were generated by the Google Cloud platform Colab [10], and the model for the last system was generated by the Phenix utility "predict_and_build" [5]. These techniques consist of two steps: structure prediction from amino acids using AlphaFold2 and model refinement using Phenix [2]. Our study used the predicted pure AlphaFold model with the highest pLDDT-score for further refinement. Then, TM-align [11] was used to align the AlphaFold model with the known true structure before the decoy generation and flexible fitting. This alignment step could alternatively have been performed by rigid-body fitting to the cryo-EM map with the *colores* or *collage* utilities of the Situs package [12], but we used the available true structure as a more direct reference frame for aligning the imperfect AlphaFold models to avoid any rigid-body fitting sources of error.

We used masked density maps generated using UCSF Chimera, as described in [5]. Specifically, we used the command "vop zone #0 #1 5 minimalBounds true modelID #2" [13], in which #0 corresponds to the cryo-EM map, and #1 represents the true structure. This function sets values to zero for grid points beyond a 5Å radius from any atom, effectively masking those areas beyond the defined zone. The use of the true structure helped eliminate additional sources of error from imperfect masking [5] that are outside the scope of this study. Then, the *map2map* utility of the Situs package [12] was used to convert the masked map to the Situs format used by ModeHunter 1.4. Other than described in this section, the true structure was used only for validation purposes (final TM-score calculation) in Sect. 2.4 below.

2.2 Decoy Generation with Normal Modes and Elastic Network Models

Normal mode analysis is widely used to study the vibrational motions of molecular structures. A normal mode is a motion pattern in which all atoms of the molecule oscillate sinusoidally at their natural frequency [14]. In a multimode superposition, every mode adds one dimension or degree of freedom to the dynamical model. Protein ENMs comprise linear (Hookean) springs connecting carbon-alpha atoms within a certain cutoff distance. The normal modes of ENMs, when used as models of protein dynamics, yield a low-dimensional collection of basic functions for the fundamental motions of the structure. Normal mode analysis was applied using the ModeHunter package [15].

To generate normal modes and frequencies from AlphaFold models, the carbon-alpha-based ENM tool of ModeHunter, *enmhunt.py*, was used, with a recommended distance cutoff of 12Å. To extract the eigenvectors (normal modes) and eigenvalues (corresponding to frequencies), the second derivative matrix of the pairwise harmonic potential energy function (Hessian matrix) was diagonalized. The ModeHunter function *augment.py* was used to interpolate the modes to all atoms after the computed modes and frequencies were saved in Python pickle files.

The ModeHunter tool *eldecoy.py* was used to generate the decoys. The method has seven parameters: the PDB name of the AlphaFold model, the pickle file containing the all-atom modes, the start and end indexes of the modes, the amplitudes of the modes (determined by the RMSD from the AlphaFold model), the number of sampling phases of each mode, and the output name. The method generates trajectories for the coordinates of sequential frames (one frame for each decoy) and saves them in PDB format. In this study, we used three non-trivial lowest-frequency modes, 7, 8, and 9, for the decoys and 10 discrete sampling steps to sample each mode as a cosine wave between Phase 0 and Phase $+\pi$, with a constant amplitude of 5Å. Thus, $10^3 = 1,000$ decoys comprise the complete set of sampling combinations.

2.3 Flexible Fitting of Decoys into Cryo-Electron Microscopy Maps

To flexibly fit the 1,000 AlphaFold decoys (generated in Sect. 2.2) into the density map, the *elforge.py* tool of ModeHunter version 1.4 was employed [5, 15]. In the tool, eight parameters are used to generate fitted structures in PDB format: the PDB name of the

AlphaFold decoys, the density maps, the resolution of the density map, the pickle file containing all-atom modes, the start and end indexes of the mode range, the zero-padding margin, and the output name.

The flexible fitting method in *elforge.py* starts by loading the essential data, including normal modes, frequencies, and the AlphaFold decoys, along with the volumetric target EM map. The fitting process utilizes the SciPy optimization functions, the goal of which is to adjust the coefficients of a linear combination of normal modes applied to the AlphaFold decoys to maximize alignment with the target EM map. This is achieved by maximizing the CC-score between the deformed structure and the target map. The fitted structure is mapped onto a voxel grid and convolved with a Gaussian kernel to match the resolution of the target map. We used the optimal parameters derived in [5] in this study, which included masked maps, mode range 1–12 (including rigid-body modes 1–6), inner product CC, and local Powell optimization.

2.4 Dimensionality Reduction and Clustering

We employed MDS to reduce the data dimensionality and to visualize structural relationships among decoy models before and after fitting. Pairwise similarities between all decoys were calculated using TM-align [11]. The resulting TM-scores were converted into a dissimilarity matrix by subtracting each score from 1, with the resulting matrix serving as input for MDS. Using the MDS function from the scikit-learn package [mds = MDS(n_components = 2, dissimilarity = 'precomputed', random_state = 42)], where $n\_components = 2$ specifies that the data should be projected into two dimensions, we reduced the high-dimensional similarity data to a two-dimensional space. The argument *dissimilarity = 'precomputed'* indicates that the input is a precomputed dissimilarity matrix; in our case, a pre-calculated dissimilarity matrix based on TM-scores. The reproducibility of the results is ensured by *random_state = 42* by setting a fixed seed for the random number generator. This transformation allowed for a clearer visual interpretation of the structures and their relative relationships in a reduced form.

To further examine groupings in the data, we applied DBSCAN to the MDS embeddings of the flexibly fitted decoys. DBSCAN is a clustering algorithm that groups points based on density. We clustered the flexibly fitted decoys using the command [dbscan = DBSCAN (eps = 0.01, min_samples = 10)], which highlighted distinct clusters based on structural dissimilarities. The parameter *eps = 0.01* defines the maximum distance between two points for them to be considered part of the same neighborhood, while $min\_samples = 10$ specifies the minimum number of points required to form a dense cluster. This approach provided a deeper understanding of the distribution of structural variations and similarities within the decoy set, facilitating both visual analysis and group-level insights of the precision of the fits.

The resulting 1,000 flexibly fitted decoys were finally compared to the true structure using TM-align [11]. The resulting TM-scores with the known true structure provide a measure of the accuracy of the fits, as shown in Figs. 1, 2, 3 and 4B and C. Moreover, the CC-scores of the optimized elongation generated from the inner product calculation (Sect. 2.3) were extracted and normalized to a range between 0 and 1 among the fitted decoys:

$$\text{Normalized CC} = x - \min(x) / \max(x) - \min(x), \quad (1)$$

where x is the CC-score to be normalized, $\min(x)$ is the minimum CC-score and $\max(x)$ is the maximum CC-score among the fitted decoys. Then, the TM-scores and normalized CC scores of each flexibly fitted decoy were plotted to investigate the global characteristics of the CC objective function as shown in Figs. 1, 2, 3 and 4C.

3 Results

For the lipid-preserved respiratory supercomplex system, the embedding of the decoy distribution into a plane using the MDS of the mutual TM-scores (Sect. 2.4) is shown in Fig. 1A before any flexible fitting. The tools *enmhunt.py* and *eldecoy.py* were used to generate the initial 1,000 decoys from the AlphaFold model (Sect. 2.2). As can be seen in Fig. 1A, the decoy sampling generates a hypercube of conformations that is embedded in the 2D plane (in the figure, an actual cube is embedded, since three modes, 7–9, were used), where every point corresponds to a decoy. Then, the flexible fitting tool *elforge.py* was employed to fit the 1,000 decoys (Sect. 2.3) into the corresponding 5Å density map EMD 30673. The fitted structures were aligned with the associated true structure chain 3 of PDB ID 7DGQ using TM-align [11] and clustered with MDS and DBSCAN (Sect. 2.4). Figure 1B shows each point corresponding to a flexibly fitted decoy. The ensemble of start decoys collapsed into the 11 main clusters (colored) and the unclustered outliers (light gray). The clusters in Fig. 1B were labeled by the best TM-score, with the black arrow indicating the best fit (corresponding to the highest TM-score among all fitted decoys). Figure 1C shows the normalized CC-score plotted against the TM-score. Each point in Fig. 1C corresponds to a fitted decoy, and the color coding corresponds to that in Fig. 1B, differentiating between the various clusters and outliers. We observed that most of the fitted decoys had TM-scores of 0.50–0.70. The high TM-scores (0.65–0.70) were associated with high normalized CC-scores (0.9–1.0), indicating that these decoys (mainly from the large blue cluster) were close to the true structure both in terms of structure alignment and CC. The spread in the middle range (TM-scores of 0.55–0.65) shows high normalized CC-scores, but not as high as in the blue cluster. Decoys with low TM-scores (< 0.55) were spread across a wide range of normalized CC-scores, showing less correlation between the two metrics in this region. As shown in Figs. 1B and C, the blue cluster stands out as the winning ensemble, achieving the highest TM-score of 0.70, which is significantly better than the 0.52 TM-score of the AlphaFold model; and the blue cluster also produced the best CC-scores we observed (Fig. 1C), indicating a good match with the experimental data. In contrast, the other clusters, such as the red cluster, represent spurious ensembles, where deformed structures were attracted to lower local maxima of the CC coefficient, leading to suboptimal fits (Figs. 1B and C).

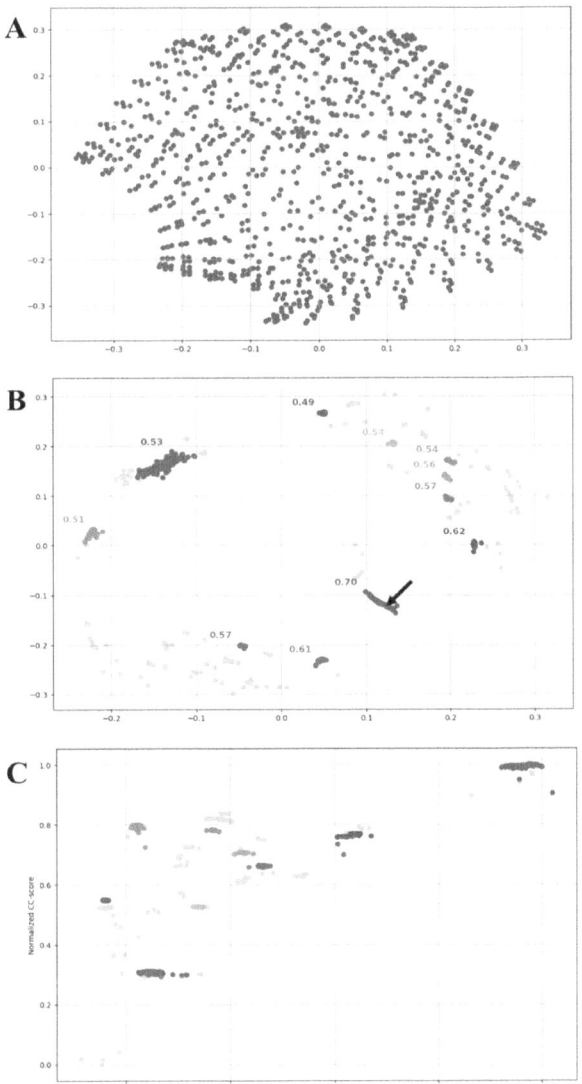

Fig. 1. Visualization of 1,000 decoys of the lipid-preserved respiratory supercomplex. Decoys were embedded in 2D principal coordinates according to their mutual TM-scores (A) before and (B) after flexible fitting. The dimensionality reduction in (A) and (B) was performed using MDS [8]. The DBSCAN algorithm was used to cluster the fitted decoys in (B) [9]. The highest TM-score among the cluster members is displayed with the corresponding color for each cluster in (B). The black arrow in (B) identifies the winning best-fit structure. (C) Comparison of the TM-score and the normalized CC-score for a set of decoys. The TM-score was calculated using TM-align [11], and our flexible fitting method generated the normalized CC-score [5].

Figure 2A shows the distribution of the initial 1,000 decoys of the flagellar L-ring protein embedded in 2D principal coordinates based on their mutual TM-scores. The decoys were created and analyzed as described in Sects. 2.2 and 2.4. The decoys of the flagellar L-ring protein system (Sect. 2.3) were flexibly fitted into the appropriate 5Å density map EMD 12183 using *elforge.py*. Following TM-align [11] with the corresponding actual structure chain A of PDB ID 7BGL, the fitted structures were clustered (Sect. 2.4) as shown in Fig. 2B. We observed 10 main clusters (colored) and unassigned outliers (light gray). The highest TM-score is shown in the corresponding color (Fig. 2B). For each of the 1,000 fitted decoys, the normalized CC-score is plotted against the TM-score in Fig. 2C, with the colors corresponding to those of the clusters in Fig. 2B. Most of the TM-scores in Fig. 2C were within 0.58–0.65 (and their normalized CC-scores above 0.5) demonstrating that the corresponding decoy fits (which are primarily from the top right corner clusters) are close to the known true structure. Decoys with low TM-scores (< 0.54) exhibited only low normalized CC-scores. In Figs. 2B and C, the blue cluster achieved the highest TM-score of 0.65, surpassing the AlphaFold model's TM-score of 0.53. This demonstrates a better match with the known true structure. Moreover, a high CC value indicates an accurate fit with the experimental data. The other clusters, such as the red, green, orange, and brown clusters, also had relatively high TM-scores and CC-scores. This suggests that the CC-scores effectively captured structural similarities across various ensembles, emphasizing the robustness of the fitting process. On the other hand, the decoys at the bottom-left corner denote structures trapped in incorrect local maxima of the CC, resulting in a poorer fit with the true structure.

The 1,000 decoys of the cation diffusion facilitator YiiP system in Fig. 3A were created and analyzed as described in Sects. 2.2 and 2.4. Then, the decoys (Fig. 3A) were flexibly fitted (Sect. 2.3) to the associated 5Å density map EMD 23093. After the fitted structures were aligned with the corresponding experimental structure chain B of PDB ID 7KZZ using TM-align [11], visualization and clustering were performed with MDS and DBSCAN (Sect. 2.4), with each point in Fig. 3B representing a flexibly fitted decoy. Flexible fitting reduced the structural differences between the decoys and, in this case, produced only one winning cluster shown in blue (light gray denotes the unassigned outliers). The single cluster contains the best TM-score result of 0.83 (Fig. 3B), indicated by the black arrow. This demonstrates that the decoys in this cluster were the most structurally similar to the known true structure after flexible fitting, thereby improving the AlphaFold model, which has a lower TM-score of 0.76. In contrast, the DBSCAN analysis yielded, apart from the winning cluster, only outliers that consisted of deformed structures that fit the density map differently but not as accurately as those in the dominant blue cluster. The presence of these outliers shows that even if there is only a single winning cluster, convergence is not guaranteed. However, the comparison with the CC in Fig. 3C indicates that the best (highest TM-score) fits can be readily identified by their CC with the experimental map.

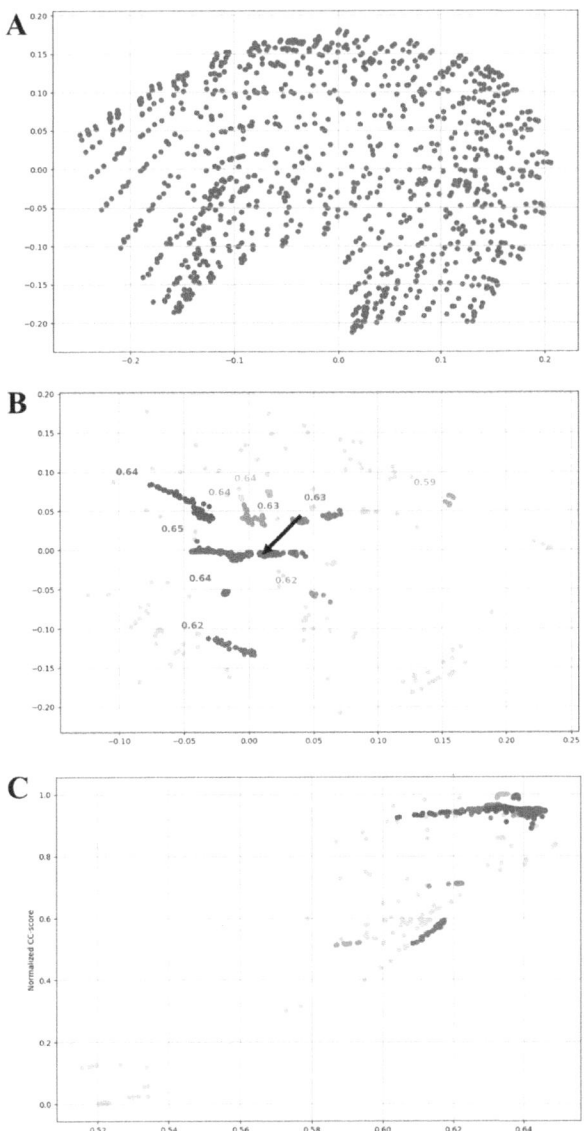

Fig. 2. Visualization of 1,000 decoys of the flagellar L-ring protein. Panels A–C were created as described in Fig. 1.

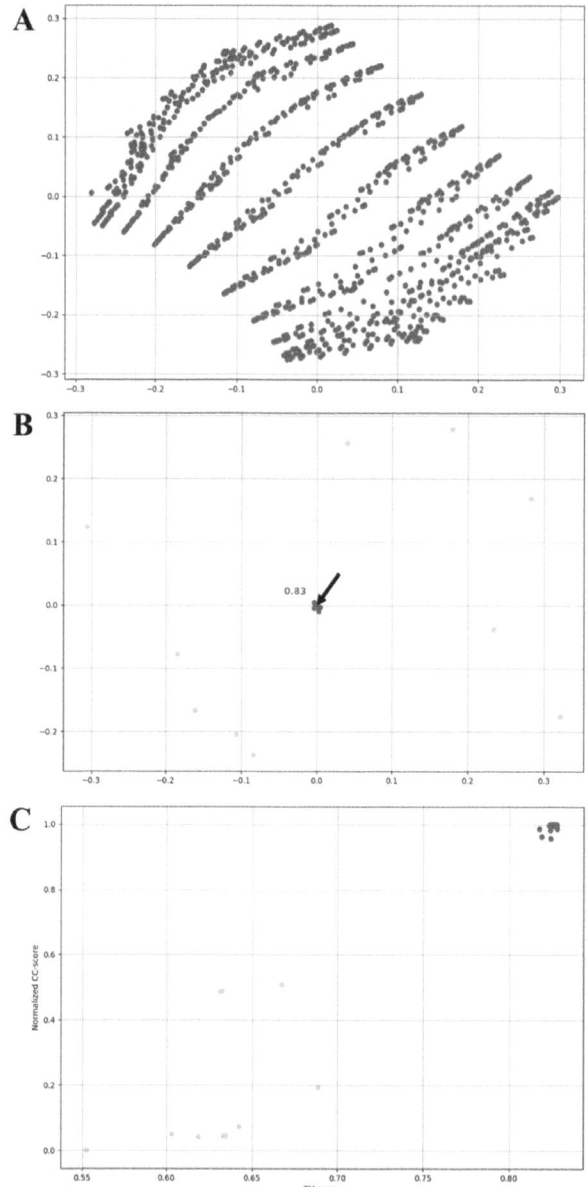

Fig. 3. Visualization of 1,000 decoys of the cation diffusion facilitator YiiP. Panels A–C were created as described in Fig. 1.

Figure 4A shows the MDS embedding of the initial decoys for the *Sulfolobus islandicus* pilus system. The decoys were created and analyzed as described in Sects. 2.2 and 2.4. Then, they were flexibly fitted into the corresponding 4.10Å density map, EMD 0397 (Sect. 2.3). After the fitted structures were aligned with the related true structure chain

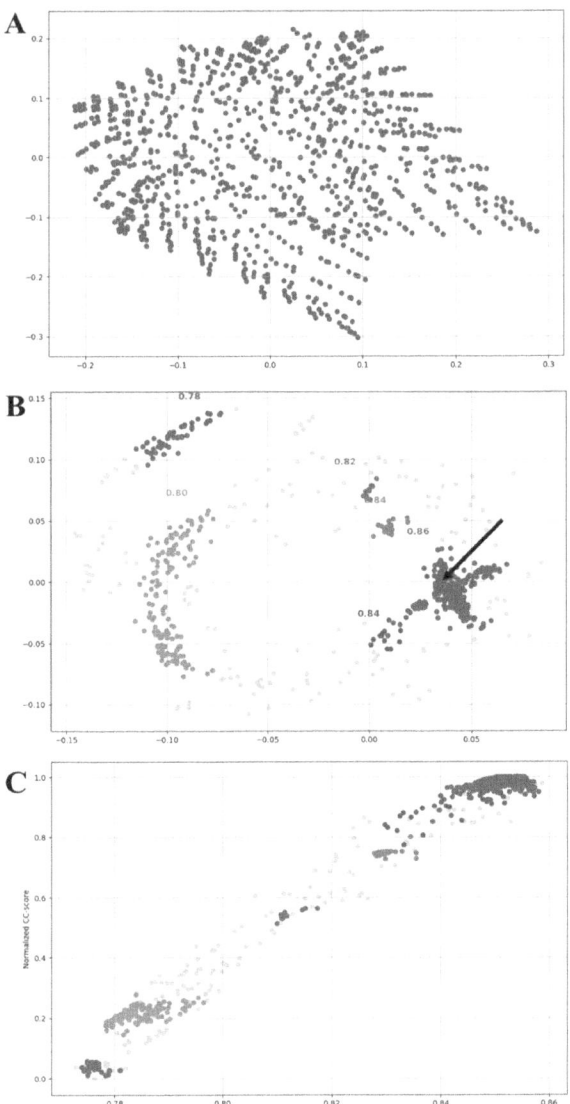

Fig. 4. Visualization of 1,000 decoys of *Sulfolobus islandicus* pilus. Panels A–C were created as described in Fig. 1.

A of PDB ID 6NAV using TM-align [11], they were submitted to MDS and DBSCAN (Sect. 2.4) for visualization and clustering (Fig. 4B). This resulted in the decoys becoming more clearly grouped into six main clusters (colored) and unassigned outliers (light gray), with each cluster's highest TM-score shown in the corresponding color (Fig. 4B). The normalized CC-scores are plotted against the TM-scores in Fig. 4C (with the colors corresponding to that of their cluster in Fig. 4B). A strong correlation was observed between the TM-scores and the CC-scores in Fig. 4C, indicating that decoys that are structurally similar to the target structure also tend to fit better into the experimental reference map. Figures 4B and C show that the blue cluster achieved the highest TM-score of 0.86, thereby improving the AlphaFold model score of 0.77. On the other hand, the other clusters, such as the red and orange clusters, represent ensembles where deformed structures were attracted to the lower local maxima of the CC, resulting in spurious fits.

4 Discussion

Our observations highlight the advantages of broader sampling of the conformational space (Figs. 1, 2, 3 and 4A) in overcoming the limitations associated with relying on a single model, which may become trapped in the local maxima of the CC during optimization (Figs. 1, 2, 3 and 4B). For instance, the decoys from the red cluster, as shown in Figs. 1 and 4, were attracted to the local maximum, resulting in an inaccurate structure, as indicated by the low TM-scores and CC-scores. Thus, employing diverse conformational ensembles enhances the accuracy of model refinement by mitigating the risks of optimization traps. Starting with a good initial structure avoids getting trapped in local maxima of the CC and results in more accurate fits (as seen in the blue cluster in Figs. 1, 2, 3 and 4B and C). This highlights that generating a reliable starting structure can significantly improve the effectiveness of the refinement process. By generating alternative structures (decoys), our method provides a more complete exploration that facilitates the identification of the best solution and of spurious fits.

Moreover, our multi-shot approach offers a better sampling in the winning cluster ensemble than with fitting a single structure, as evidenced by the blue cluster with the highest TM-score and CC-score in Figs. 1, 2, 3 and 4B and C. For the first system (Fig. 1), the TM-score increased from 0.52 to 0.70 with multi-shot, compared to only 0.69 achieved in one-shot refinement [5], Similarly, for the second system (Fig. 2), the TM-score rose from 0.53 to 0.65 in multi-shot, compared to 0.63 in one-shot [5]. For the fourth system (Fig. 4), the TM-score rose from 0.77 to 0.86 in multi-shot, compared to 0.85 in one-shot [5]. These results demonstrate that our multi-shot method can theoretically yield slightly better structural refinements than the earlier one-shot strategies if these optimum structures can be identified among the winning cluster members.

As shown in Figs. 1, 2, 3 and 4C, relative CC-scores are a good predictor of accuracy, as evidenced by the correlation between the TM-scores (which are based on the known true structures) and the CC-scores (which are based on the match with the cryo-EM maps). The CC scores in Figs. 1, 3 and 4 can be used to identify the winning cluster. However, the ranking of custers by the CC in Fig. 2C (top right) is complicated by minor discrepancies between an experimental map and the true structure. Moreover, TM-scores for the highest CC scores among the four cases (Figs. 1, 2, 3 and 4) were 0.69, 0.63, 0.83, and 0.85, respectively (identical to the scores achieved by one-shot refinement [5]). Therefore, further studies of the winning cluster ensemble and of appropriate screening functions are required to take advantage of the better sampling afforded by the multi-shot approach.

Although many start structures need to be evaluated in the multi-shot approach, the use of the efficient Powell optimization (about one order of magnitude faster than global methods [5]) can offset this computational cost. For example, in [5], the highest accuracy results achieved in the first and fourth systems by using 8–10 times slower global optimization methods were the same as those achieved with Powell (TM-scores: 0.69 and 0.85, respectively).

5 Conclusion

The observation of multiple basins of attraction and their corresponding solution clusters in Figs. 1, 2, and 4 represents a paradigm shift in cryo-EM flexible fitting. Earlier studies on cryo-EM flexible fitting by other groups [16, 17], developed before the 2014 resolution revolution [18] of cryo-EM, was based on the "one start structure, one best fit" paradigm. It is likely that the low resolution of the maps prior to 2014 created a smoother CC landscape that avoided local maxima. This hypothesis could be tested in future studies through systematic lowering of the resolution in the aforementioned cases (Figs. 1, 2, and 4) using hybrid maps [3] to see if such cases would become more similar to that in Fig. 3 that exhibited only a single dominant cluster. Our results also call for more detailed characterization of the winning cluster ensemble (since the "correct fit" is no longer a single structure). Specifically, it is worth exploring whether the observed benefit of using decoys (leading to slightly improved fits over single start structures) can be used in practical applications (e.g., for screening the winning cluster ensemble with a suitable scoring function that detects such minor improvements).

Acknowledgments. We thank Min Dong for providing IT support for the software installation and Jiangwen Sun for the discussions in the early phases of this study.

Funding. This work was supported by National Institutes of Health Grants R01-GM062968 and R35-GM153431, the Old Dominion University Batten Endowment to W.W., and a scholarship for M.A. from the Government of Saudi Arabia.

Data Availability. Version 1.4 of the software ModeHunter can be freely downloaded at https://modehunter.biomachina.org.

Disclosure of Interests. The authors have no competing interests to declare that are relevant to the content of this article.

References

1. Jumper, J., et al.: Highly accurate protein structure prediction with AlphaFold. Nature **596**(7873), 583–589 (2021)
2. Terwilliger, T.C., et al.: Improving AlphaFold modeling using implicit information from experimental density maps. bioRxiv (2022)
3. Alshammari, M., et al.: Refinement of AlphaFold2 models against experimental and hybrid Cryo-EM density maps. QRB Discovery **3**, 1–23 (2022)
4. Alshammari, M., He, J., Wriggers, W.: Refinement of AlphaFold2 models against experimental Cryo-EM density maps at 4–6Å resolution. In: 2022 IEEE International Conference on Bioinformatics and Biomedicine (BIBM), pp. 3423–3430. IEEE (2022)

5. Alshammari, M., He, J., Wriggers, W.: Flexible fitting of alphafold2-predicted models to Cryo-EM density maps using elastic network models: a methodical affirmation. Bioinform. Adv. **5**(1), vbae181 (2025). https://doi.org/10.1093/bioadv/vbae181
6. Cui, Q., Bahar, I.: Normal mode analysis: theory and applications to biological and chemical systems. CRC Press (2005)
7. Alshammari, M., He, J., Wriggers, W.: AlphaFold2 model refinement using structure decoys. In: Proceedings of the 14th ACM International Conference on Bioinformatics, Computational Biology, and Health Informatics, pp. 1–7 (2023)
8. MDS. https://scikit-learn.org/dev/modules/generated/sklearn.manifold.MDS.html. Accessed 15 Oct 2024
9. Density-Based Spatial Clustering of Applications with Noise, https://scikit-learn.org/dev/modules/generated/sklearn.cluster.DBSCAN.html, last accessed 10/15/2024
10. AlphaFold with a density map. https://colab.research.google.com/github/phenix-project/Colabs/blob/main/alphafold2/AlphaFoldWithDensityMap.ipynb. Accessed 16 May 2022
11. Zhang, Y., Skolnick, J.: TM-align: a protein structure alignment algorithm based on the TM-score. Nucleic Acids Res. **33**(7), 2302–2309 (2005)
12. Wriggers, W.: Conventions and workflows for using Situs. Acta Crystallogr. D Biol. Crystallogr. **68**(4), 344–351 (2012)
13. Pettersen, E.F., et al.: UCSF Chimera—A visualization system for exploratory research and analysis. J. Comput. Chem. **25**(13), 1605–1612 (2004)
14. Normal mode. https://en.wikipedia.org/wiki/Normal_mode. Accessed 10 July 2023
15. Wriggers, W., et al.: ModeHunter: a package for reductionist analysis, animation, and application of elastic biomolecular motion. J. Phys. Chem. B **129**(3), 825–834 (2025). https://doi.org/10.1021/acs.jpcb.4c05077
16. Tama, F., Miyashita, O., Brooks, C.L., 3rd.: Normal mode based flexible fitting of high-resolution structure into low-resolution experimental data from cryo-EM. J. Struct. Biol. **147**(3), 315–326 (2004)
17. López-Blanco, J.R., Chacón, P.: IMODFIT: efficient and robust flexible fitting based on vibrational analysis in internal coordinates. J. Struct. Biol. **184**(2), 261–270 (2013)
18. Kühlbrandt, W.: The resolution revolution. Science **343**(6178), 1443–1444 (2014)

Using Autoencoders to Explore the Conformational Space of the Cdc42 Protein

Fatemeh Afrasiabi, Ramin Dehghanpoor, and Nurit Haspel(✉)

University of Massachusetts Boston, Boston, MA 02125, USA
nurit.haspel@umb.edu

Abstract. Understanding protein structure and dynamics is essential for understanding their function. This is a challenging task due to the high complexity of the conformational landscapes of proteins and their rugged energy levels. In particular, it is important to detect highly populated regions which could correspond to intermediate structures or local minima. In this work we train a neural network model on the MD simulations data to create a low-dimensional latent space. The latent space is further used to explore the protein's conformational space. It can be used to be interpolated or extrapolated to produce new intermediate protein conformations that might not have been previously seen. The latent space visualization also assists in visualizing the conformational path of the respective proteins. We compare the performance of a linear autoencoder and a variational autoencoder and discuss their advantages and shortcomings for exploring the pathways of the Cdc42 protein.

Keywords: Autoencoders · Protein conformational search · Cdc42 · MD simulations

1 Introduction

Understanding protein structure and dynamics is essential for understanding their function. This is a challenging task due to the high complexity of the conformational landscapes of proteins and their rugged energy levels. Researchers have come up with ways to explore this space with both experimental and computational methods and that each of these methods has specific limitations either in accuracy or computational complexity. One popular physics-based computational method is Molecular Dynamics (MD) Simulations [20]. In classical MD simulations, Newton's equations of motion are utilized to calculate how each atom in a molecule changes its position, velocity, and acceleration in a computer simulation environment. MD simulations predict how an interacting system evolves in time by generating potential atomic trajectories of the system. The calculation of forces acting on the atoms and their potential energies typically involves the use of inter-atomic potentials or molecular mechanics force fields. MD simulations have emerged as a popular method for studying the kinetic behavior of proteins [15,16,25]. This approach can effectively represent a system's potential energy landscape and sample the conformational spaces of the protein. Given the complex and dynamic nature of proteins and their interactions with molecules in the environment, the ability to simulate these interactions using MD simulations is remarkably valuable [21,22]. Experimental methods like X-ray crystallography [11,19] have been and are currently used to discover 3D structures. Still, they only provide snapshots and cannot capture the full scale of conformational changes, especially under specific physiological and environmental conditions. These snapshots can be used as starting points for MD simulations where inaccessible time scales can be simulated with high levels of accuracy. MD simulations can be used to simulate time scales ranging from femtoseconds to microseconds but still are incapable of capturing longer time scales due to the high computational complexity of the simulations, even on state-of-the-art computing servers. As standard MD simulations fall short of simulating more complex and timely dynamics, researchers have worked on variations of MD simulations to enable capturing greater timescales such as Steered MD [17], Replica Exchange MD [26], Targeted MD [31], and more. Researchers have developed and applied machine-learning-based methods to explore the conformational space of proteins [9,28].

Autoencoders [4] are an unsupervised neural network model. They consist of an encoder where the input data is mapped to lower-dimensional representations and a decoder part where the lower-dimensional representations, often called the latent space or the bottleneck, are mapped back to their original representation. Mapping the latent space back to the original dimension is also called reconstruction as the autoencoder tries to reconstruct the original data with the least possible error. The autoencoder is trained with the minimization of the reconstruction loss as the objective function. The architecture of the autoencoders varies depending on how the encoder and decoder hidden layers and the latent space are designed, ranging from basic linear autoencoders to deep convolutional or recurrent autoencoders. Autoencoders are used for Dimensionality Reduction, Generative Models Pretraining, data compression and more. In the past, we used autoencoders to classify protein families with high accuracy [10]. In this work we also use a Variational Autoencoder (VAE) [18], a generative model that can be used for sampling new data points from the learned distribution. VAE adds a regularization term to the objective function, which encourages the latent space to have a (typically) normal distribution. This allows the model to generate new protein conformations that follow the training data distribution while also ensuring that the generated conformations are physically credible. Additionally, the VAE model provides a way to quantify the similarity between protein conformations by measuring the distance in the latent space. This can be useful in understanding the conformational changes occurring during protein dynamics and exploring the association between protein structure and function. Using autoencoders on MD simulation data provides a way to explore the conformational space of proteins and generate new feasible protein conformations.

In our previous work [1,2], we introduced and evaluated an algorithm that simulates the protein conformational space called the RRTMC model. Although RRTMC uses potential energy thresholds from the amber force field [29], there is still the possibility of generating intermediate conformations that are not physically plausible due to the randomness involved in generating new structures. In this work we use MD simulation data to analyze and produce physically possible conformations. We train a neural network model on the MD simulations data to create a low-dimensional latent space. The latent space is further used to explore the protein's conformational space. It can be used to be interpolated or extrapolated to produce new intermediate protein conformations that might not have been previously seen. The latent space visualization also assists in visualizing the conformational path of the respective proteins. We compare the performance of a linear autoencoder and a variational autoencoder and discuss their advantages and shortcomings for exploring the pathways of Cdc42.

2 Materials and Methods

Cell division control protein 42 homolog (Cdc42) is a Ras-related GTP-binding protein. It is a member of the Rho subfamily GTPases that act as molecular switches between active GTP-bound and inactive GDP-bound forms. The active phase (GTP-bound) of this protein interacts with downstream effectors to regulate cellular processes [7,12,30]. Cdc42 plays an essential role in several processes, such as cell cycle progression, cell signaling, and many more [6,24]. Hence, it has been shown that the dysregulation of the Cdc42 protein function can be the cause of cancer metastasis and progression [3,27].

2.1 MD Simulation Details

The Cdc42 proteins in GDP-bound and GTP-bound states were modeled using the crystal structures from the Protein Data Bank (PDB: 4DID (chain A) and PDB: 4JS0 (chain A), respectively). We substituted the GNP molecule in 4JS0 with GTP and removed the attached inhibitor. CHARMM [5] was employed to add missing side chains and hydrogen atoms. The G12V and Q61L mutants were also modeled with CHARMM, using the wild-type system as a reference. Simulations were performed using the CHARMM force field, and the TIP3P model was applied for explicit solvent representation.

Simulations were conducted using the NAMD 2.13 software package [23] and followed a previously published protocol [14]. The process began with potential energy minimization using 10,000 conjugate gradient steps. Subsequently, protein atoms were held fixed while the solvent was heated to 310 K to ensure uniform ion distribution in the solution. The system was then equilibrated isothermally and isobarically at 310 K and 1 bar (NPT conditions) for 500 ps to reach infinite dilution conditions (i.e., water density close to 1 g/cm3). The solute was allowed to move, and the entire system was heated and equilibrated (50 ps) at the production temperature of 310 K and pressure of 1 bar. A numerical integration time step of 2 fs was employed for all simulations, and the nonbonded pair list was updated every five steps. During production runs, the constant temperature of 310 K was maintained by Langevin temperature control, and the pressure at 1 atm was sustained by Nosé-Hoover Langevin piston pressure control. Trajectories were recorded every 50,000 steps (25 ps) for subsequent analysis during the production runs, with each simulation running for 500 ns for six systems, totaling 3 μs. Full details about the simulations can be found in [14].

2.2 Autoencoders

In this work, we utilized two versions of autoencoders, the basic linear autoencoder, and the variational autoencoder, to explore the protein conformational space of Cdc42 MD simulations. We show how these two types of autoencoders provide a distribution in the latent space for the Wild Type (WT), G12V (GV) mutant, and the Q61L (QL) mutant for both the active GTP and inactive GDP-bounds. Each input data point to the models is the flattened tensor of all the coordinates of the backbone atoms of each protein variant. The Cdc42 protein has a total of 709 backbone atoms; therefore, each input tensor is a float vector of length 2127. Each MD simulation contains 5000 conformations, which makes the MD data of each model trained individually for a specific protein, a matrix tensor of size 5000×2127. Figure 1(a) shows the design of our linear autoencoder. It consists of two hidden linear layers in the encoder part and two corresponding linear layers in the decoder. Each layer is followed by the ReLU activation function, batch normalization, and dropout with a probability of 20%. We experimented with the latent length of sizes two and three.

Batch normalization is a technique used to improve the training stability and subsequently the speed of the neural networks by normalizing the input data to each layer of the network while ensuring that the mean and variance of the data are consistent across each mini-batch of samples. This process helps prevent the internal covariant shift issue that may occur during training where the distribution of the input data to each layer could shift and affect the network's performance negatively. We used the ReLU (Rectified Linear Unit) activation function by setting all negative values in the input data to zero whilst leaving positive values unchanged. This can help improve the nonlinearity of the network and preclude the vanishing gradient problem that can happen when using other activation functions such as sigmoid and hyperbolic tangent (tanh).

Our variational autoencoder structure is depicted in Fig. 1(b), where we added the *mean* and *sigma* layers. The output of these layers is used to create the latent space distribution from which a sample point is next randomly chosen. ϵ is a random noise that is sampled from a standard normal distribution $\epsilon \in \mathcal{N}(0,1)$ during the decoding process. The VAE also consists of two hidden layers, followed by ReLU activation, batch normalization, and dropout layers. The Sigmoid activation function follows the last layer of the decoder. Both models were trained, validated, and tested on an 80-10-10 split of the data. The training process hyperparameters are as follows: We used a batch size of 128,

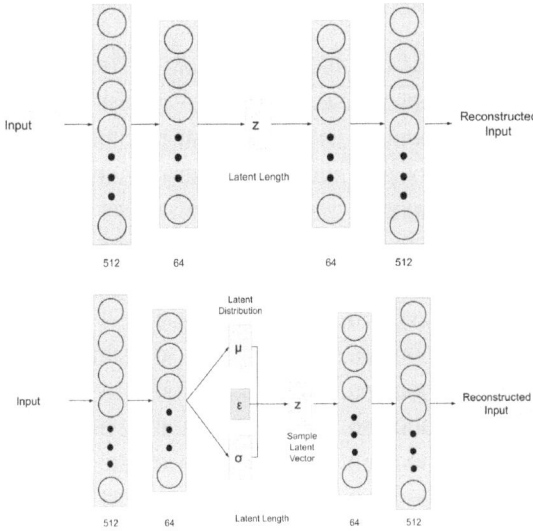

Fig. 1. (a) Linear Autoencoder trained on MD simulation data. (b) Variational Autoencoder trained on MD simulation data.

the Adam optimization algorithm and a weight decay of 0.0001 We also tested the code on different values for the number of epochs, latent length, random seed, and learning rate values. The autoencoders were developed using Python 3.9.13 [13] and PyTorch version 1.12.1 [8].

2.3 CNN Valid Conformation Classifier

In order to guide the autoencoder to generate better conformations that are physically plausible, a CNN classifier was also added that classifies conformations as valid or invalid. This classifier was trained on conformations from the MD simulation (valid conformations) and infeasible conformations that were curated to have either clashes between the atoms or distances that are not physically possible. Figure 2 shows the design of the CNN classifier. During the training of the autoencoders, when a new conformation is reconstructed from the latent space, either by sampling from the latent space distribution in the case of Variational Autoencoder or by simply reconstructing from the latent space directly, it will go through the CNN classifier and a penalty will be added to the loss values if the conformation is invalid. Figure 3(a) shows an example of a valid conformation that does not have any clashes between the atoms, and the energy value is below our set threshold. An invalid protein conformation for Cdc42 is shown in Fig. 3(b), where there are severe clashes between atoms. This example is chosen to show an extreme case of infeasible backbone coordinates. However, the CNN classifier is also trained on many artificially created infeasible conformations with different energy values and atom coordinates. This classification is

Table 1. Results on individually trained AEs and VAEs

Protein	Latent	Loss†	Silhouette	Pearson	Spearman	Kendall	MMD	EMD
GV GDP (AE)	2	**0.00027**	**0.4524**	0.4642	0.4883	0.3492	0.0838	0.0403
	3	**0.00027**	0.3806	**0.6359**	**0.6538**	**0.4718**	**0.0361**	**0.0370**
GV GDP (VAE)	2	0.00029	**0.3326**	**0.2939**	**0.3590**	**0.2428**	0.0609	0.0420
	3	**0.00028**	0.2256	0.2707	0.3260	0.2202	**0.0332**	**0.0410**
GV GTP (AE)	2	**0.00026**	**0.5739**	0.3137	0.2772	0.1895	0.2602	0.0354
	3	**0.00026**	0.2607	**0.5183**	**0.4679**	**0.3235**	**0.0629**	**0.0284**
GV GTP (VAE)	2	0.00028	**0.2870**	**0.2598**	0.2438	0.1640	0.1174	0.0316
	3	**0.00027**	0.2042	0.2510	**0.2606**	**0.1757**	**0.0660**	**0.0301**
QL GDP (AE)	2	**0.00027**	**0.5647**	0.4607	0.5319	0.3738	0.0843	0.0287
	3	**0.00027**	0.4209	**0.7091**	**0.7216**	**0.5317**	**0.0447**	**0.0272**
QL GDP (VAE)	2	**0.00028**	**0.3359**	**0.3038**	**0.3393**	**0.2293**	0.0885	0.0320
	3	**0.00028**	0.2356	0.2859	0.3225	0.2183	**0.0584**	**0.0307**
QL GTP (AE)	2	**0.00028**	**0.6317**	0.6401	0.7059	0.5100	0.1836	0.0313
	3	**0.00028**	0.4951	**0.7239**	**0.7599**	**0.5592**	**0.1033**	**0.0294**
QL GTP (VAE)	2	**0.00029**	**0.3721**	**0.3144**	**0.4014**	**0.2711**	0.1073	0.0337
	3	**0.00029**	0.2749	0.2774	0.3728	0.2544	**0.0594**	**0.0315**
WT GDP (AE)	2	**0.00027**	**0.4494**	0.4285	0.4633	0.3248	0.0984	0.0376
	3	**0.00027**	0.3296	**0.5970**	**0.6239**	**0.4442**	**0.0338**	**0.0334**
WT GDP (VAE)	2	**0.00028**	**0.3259**	0.2211	0.2674	0.1805	0.0797	0.0399
	3	**0.00028**	0.2366	**0.2401**	**0.2753**	**0.1866**	**0.0374**	**0.0377**
WT GTP (AE)	2	**0.00027**	**0.4436**	0.4668	0.4095	0.2839	0.1216	0.0321
	3	**0.00027**	0.3474	**0.6783**	**0.5494**	**0.3897**	**0.0679**	**0.0296**
WT GTP (VAE)	2	**0.00028**	**0.3041**	**0.3775**	0.3058	0.2076	0.0989	0.0316
	3	**0.00028**	0.1926	0.3598	**0.3190**	**0.2168**	**0.0562**	**0.0302**

† Loss refers to MSE measurement on the test set. For each protein, the bold numbers show better performance when comparing latent space lengths of two and three. The simulations were done using four random seed initializations and the results are averaged. For more detailed overview of the metrics used, please refer to Sect. 3.1.

especially important when a variational autoencoder is used and new conformations are generated.

3 Results

3.1 Evaluation Metrics

We evaluate the performance of our linear autoencoder and variational autoencoder with the following metrics. The distances we measured were Euclidean distances.

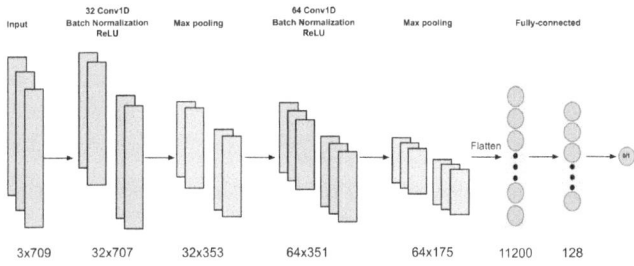

Fig. 2. CNN classifier that gets coordinates of a protein conformation and classifies them as valid or invalid in terms of potential energy.

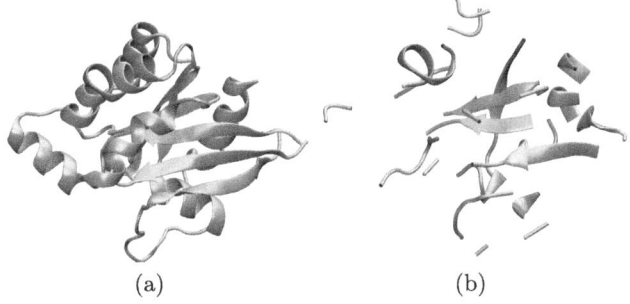

Fig. 3. (a) An example of a conformation where the CNN energy classifier classifies as **valid**. (b) An example of a conformation where the CNN energy classifier classifies as **invalid**. This conformation clearly had unfeasible angles, distances, and torsion. As a matter of fact, the structure is completely broken.

Spearman's Rank Correlation Coefficient: Spearman's rank correlation coefficient (referred to as *Spearman r*) evaluates the correlation between the Euclidean distances in the latent space and the original data. This helps us analyze how well the distances between all pairs of points in the original high-dimensional input space have been preserved in the reduced-dimensional latent space:

$$\rho = 1 - \frac{6 \sum d_i^2}{n(n^2 - 1)}$$

where ρ is the *Spearman r*, d_i is the difference in ranks between the two variables, and n is the total number of samples. The value of ρ ranges from -1 to $+1$, where -1 indicates a perfect negative correlation, $+1$ indicates a perfect positive correlation, and 0 indicates no correlation.

The Pearson Correlation Coefficient: Pearson correlation coefficient (referred to as *Pearson r*) is a number between -1 and 1 that measures the strength of the linear relationship between two variables. We use *Pearson r* to evaluate the correlation between the distances in the latent space and the distances between the original data:

$$r_s = 1 - \frac{6 \sum d_i^2}{n(n^2 - 1)}$$

where d_i is the difference between the ranks of the corresponding pairs of variables and n is the sample size.

Kendall's Rank Correlation Coefficient: Also called Kendall's Tau, (referred to as *Kendall r*) is a nonparametric method measure of relationships between columns of ranked data. In this study, we use Kendall's rank correlation coefficient to evaluate the correlation between the distances in the latent space and the distances between the original data and it is calculated by

$$\tau = \frac{\text{(No. concordant pairs)} - \text{(No. discordant pairs)}}{\binom{n}{2}}$$

Concordant pairs are the data points that have the same ranking order in both variables where we want to calculate the correlation.

Silhouette Coefficient: A metric to evaluate the quality of clustering. It ranges from -1 to 1 where a higher value indicates better clustering with more coherent clusters. The silhouette coefficient for the ith data point is calculated as:

$$s_i = \frac{b_i - a_i}{\max a_i, b_i}$$

where a_i is the average distance between the ith data point and all others in the same cluster, and b_i is the average distance between the ith data point and all data points in the nearest cluster to which the ith data point does not belong. In this study, we used the silhouette coefficient to evaluate the quality of the clustering in the latent space.

Maximum Mean Discrepancy (MMD): A measure of the difference between two probability distributions. It evaluates the distance between the empirical distribution of the data and the model's distribution in the latent space. We used the MMD to evaluate the goodness of fit of the latent space distribution to the true distribution of the protein conformations.

$$MMD^2(P,Q) = ||\frac{1}{n}\sum_{i=1}^{n} \phi(x_i) - \frac{1}{m}\sum_{j=1}^{m} \phi(y_j)||^2$$

Where P and Q are two probability distributions over space X, x_i values are samples from distribution P and y_i values from distribution Q and ϕ is a feature map that maps elements of X to a reproducing kernel Hilbert space (RKHS).

Earth Mover's Distance (EMD): Also called the Wasserstein Metric, is a measure of the distance between two probability distributions. It evaluates the minimum amount of work needed to transform one distribution into the other. In this study, we used the EMD to evaluate the distance between the latent space distribution and the true distribution of the protein conformations.

$$EMD(P,Q) = \min_{\gamma \in \Gamma(P,Q)} \sum_{i,j} c_{ij} \gamma_{ij}$$

where c_{ij} is the cost of moving a unit of mass from x_i in distribution P to y_j in distribution Q, γ_{ij} is the amount of mass moved from x_i to y_j, and $\Gamma(P,Q)$ is the set of all possible mass transport plans that move the mass from P to Q (i.e., $\gamma_{ij} >= 0, \sum_i \gamma_{ij} = Q(y_j), \sum_j \gamma_{ij} = P(x_i)$).

3.2 AE and VAE Latent Spaces

Figure 4 shows the structure of our variational autoencoder and the resulting 2D latent space for one of our example proteins, the G12V mutation of the GDP-bound protein. A conformation was randomly chosen from the original input space, shown in blue using VMD - New Cartoon, and the corresponding reconstructed structure is shown in red with the lRMSD value of 1.23Å.

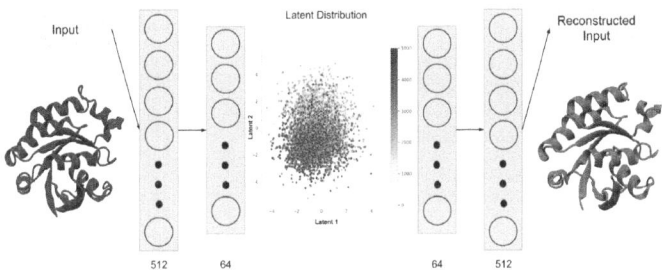

Fig. 4. Variational Autoencoder on G12V mutation of GDP-bound protein with a 2D latent space. The original conformation is shown in blue, and the reconstructed conformation is shown in red. (Color figure online)

Figure 5 depicts the 2D latent spaces of the G12V mutation, for the GDP-bound and GTP-bound proteins, created using AE and VAE. The x and y axes are the values for the 2 dimensions of the 2D latent space. The data points are colored based on their MD simulation order, i.e. 0 is the starting conformation and 4999 is the last conformation in the simulation. We can see that the conformations that happen consecutively in the MD simulation process, also appear close in the latent spaces of the AE and the VAEs. As shown in the Figure, GDP-bound simulations are more variant than GTP-bound simulations in both the latent space plots of the AE and the VAE. This is a known property of the Cdc42 molecule [14]. It is interesting to see that the VAE has created a distribution for all the possible conformations (5(a) and 5(b)), while the AE simply shows the conformational path obtained by our simulation (5(c) and 5(d)). We can use the VAE latent space from which to sample new data points and generate new conformations by reconstructing these new 2D points.

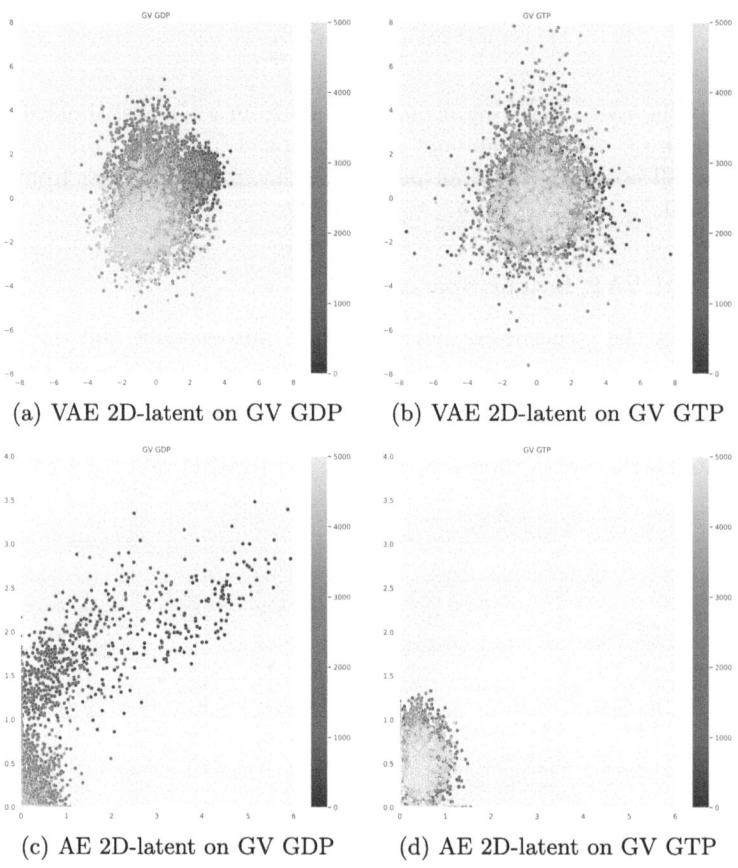

Fig. 5. AE/VAE for the G12V mutation. The x-axis shows the first dimension values of the 2D latent (latent 1) and the y-axis shows the second dimension values (latent 2). Data points are colored based on their order in the simulation process.

Table 1 shows the results of individually trained **AEs** and **VAEs** on Cdc42 protein variants with latent space dimensionality of 2 and 3. Each model and latent space combination has been trained with four random seed initializations and the results are averaged. The results illustrate that the performance of AEs varies among different proteins and latent space dimensions. For each model and evaluation metric, the one with the superior score is shown by the bold numbers in the table. In most cases, AEs with a three-dimensional latent space outperformed those with a two-dimensional latent space. We hypothesize that this is due to the fact that two dimensions may not be enough to capture the variance in the data, and higher dimensions can achieve better correlation coefficients. We can observe in Table 1 that the performance of the VAEs varies across different proteins for different latent space dimensions. For instance, for the GV GPD-bound protein, the VAE trained with a latent space dimensionality of two had better Silhouette, Pearson r, Spearman r, and Kendall r scores compared to the VAE with a latent space dimensionality of three, although the MMD and EMD scores were weaker. Overall, for most proteins, the VAEs with a latent space dimensionality of two outperformed those with a 3D latent space, however, in the few cases where this was not the case, the performance scores were very similar.

4 Conclusions

In this study, we adopt a new approach toward exploring the conformational space of proteins where we utilize MD simulation data that possess higher resolution than other computational methods and generate physically feasible conformations. We develop a neural network model that can effectively learn from the MD simulation data and construct a low-dimensional latent space. The resultant latent space can then be used to explore the protein's conformational space and at the same time allow the creation of new intermediate protein conformations that were previously unknown. The visualization of the latent space with the basic linear autoencoder enabled us to view the conformational path of the respective proteins. By comparing the performance of a linear autoencoder and a variational autoencoder, we assess the advantages and limitations of each for investigating the pathways of Cdc42. Our results show that the VAE can create a distribution of possible conformations, which can be used to generate new conformations. The 2D and 3D latent space plots generated by the VAE and the AE provided insights into the conformational changes of the Cdc42 protein variants. Our results also reveal that the GDP-bound simulations show more variance than the GTP-bound simulations, which is supported by our previous work. We also showed that for AEs latent space of three dimensions produces better scores and for the VAEs, the ones with a latent space of two are generally more effective in capturing the features of the data. In summary, our findings manifest the potential of using the variants of autoencoders to model and analyze the conformational space of proteins. In future work, we plan to interpolate and extrapolate new data points from the latent space and generate new conformational pathways that can be then proved by MD simulations.

Acknowledgments. The Molecular Dynamics simulations were run on the UMass Boston's research cluster.

Disclosure of Interests. The authors have no competing interests to declare that are relevant to the content of this article.

References

1. Afrasiabi, F., Dehghanpoor, R., Haspel, N.: Integrating rigidity analysis into the exploration of protein conformational pathways using RRT* and MC (2021). https://doi.org/10.3390/molecules26082329
2. Afrasiabi, F., Haspel, N.: Efficient exploration of protein conformational pathways using RRT* and MC. In: Proceedings of the 11th ACM International Conference on Bioinformatics, Computational Biology and Health Informatics. BCB 2020, Association for Computing Machinery, New York, NY, USA (2020). https://doi.org/10.1145/3388440.3414705
3. Aspenström, P.: Activated rho GTPASES in cancer—The beginning of a new paradigm (2018). https://doi.org/10.3390/ijms19123949
4. Baldi, P.: Autoencoders, unsupervised learning, and deep architectures. In: Guyon, I., Dror, G., Lemaire, V., Taylor, G., Silver, D. (eds.) Proceedings of ICML Workshop on Unsupervised and Transfer Learning. Proceedings of Machine Learning Research, vol. 27, pp. 37–49. PMLR, Bellevue, Washington, USA (2012). https://proceedings.mlr.press/v27/baldi12a.html
5. Brooks, B.R., Bruccoleri, R.E., Olafson, B.D., States, D.J., Swaminathan, S., Karplus, M.: CHARMM: a program for macromolecular energy, minimization, and dynamics calculations (1983). https://doi.org/10.1002/jcc.540040211
6. Bustelo, X.R.: Rho GTPASES in cancer: known facts, open questions, and therapeutic challenges (2018). https://doi.org/10.1042/bst20170531
7. Chou, M.M., Blenis, J.: The 70 KDA S6 kinase complexes with and is activated by the rho family g proteins CDC42 and RAC1 (1996). https://doi.org/10.1016/s0092-8674(00)81257-x
8. Paszke, A., et al.: PyTorch: An imperative style, high-performance deep learning library. Adv. Neural Inf. Process. Syst. **32**, 8024–8035 (2019). https://pytorch.org/docs/1.12/
9. Degiacomi, M.T.: Coupling molecular dynamics and deep learning to mine protein conformational space. Structure **27**(6), 1034–1040 (2019)
10. Dehghanpoor, R., et al.: Classifying protein families with learned compressed representations. In: Proceedings of BICOB (International conference on Bioinformatics and Computational Biology), vol. 92, pp. 47–57 (2023)
11. Drenth, J.: Principles of Protein X-ray Crystallography. Springer Science & Business Media (2007). https://doi.org/10.1007/0-387-33746-6
12. Erickson, J.W., Cerione, R.A.: Multiple roles for CDC42 in cell regulation. Curr. Opin. Cell Biol. **13**(2), 153–157 (2001)
13. Foundation, P.S.: Python language reference (2020). https://www.python.org/downloads/release/python-3913/
14. Haspel, N., Jang, H., Nussinov, R.: Active and inactive CDC42 differ in their insert region conformational dynamics (2021). https://doi.org/10.1016/j.bpj.2020.12.007
15. Hernández-Rodríguez, M., C Rosales-Hernández, M., E Mendieta-Wejebe, J., Martínez-Archundia, M., Correa Basurto, J.: Current tools and methods in Molecular Dynamics (MD) simulations for drug design. Curr. Med. Chem. **23**(34), 3909–3924 (2016)
16. Hospital, A., Goñi, J.R., Orozco, M., Gelpí, J.: Molecular dynamics simulations: advances and applications. In: Advances and Applications in Bioinformatics and Chemistry, pp. 37–47 (2015)
17. Izrailev, S., Stepaniants, S., Isralewitz, B., Kosztin, D., Lu, H., Molnar, F., Wriggers, W., Schulten, K.: Steered molecular dynamics. In: Deuflhard, P., Hermans, J.,

Leimkuhler, B., Mark, A.E., Reich, S., Skeel, R.D. (eds.) Computational Molecular Dynamics: Challenges, Methods, Ideas, pp. 39–65. Springer, Berlin Heidelberg, Berlin, Heidelberg (1999). https://doi.org/10.1007/978-3-642-58360-5_2
18. Kingma, D.P., Welling, M.: Auto-encoding variational bayes. CoRR abs/1312.6114 (2013). https://arxiv.org/abs/1312.6114
19. Maveyraud, L., Mourey, L.: Protein x-ray crystallography and drug discovery. Molecules **25**(5), 1030 (2020). https://doi.org/10.3390/molecules25051030
20. McCammon, J.A., Gelin, B.R., Karplus, M.: Dynamics of folded proteins (1977). https://doi.org/10.1038/267585a0
21. Meinhold, L., Smith, J.C., Kitao, A., Zewail, A.H.: Picosecond fluctuating protein energy landscape mapped by pressure-temperature molecular dynamics simulation. Proc. Natl. Acad. Sci. **104**(44), 17261–17265 (2007)
22. Papaleo, E., Mereghetti, P., Fantucci, P., Grandori, R., De Gioia, L.: Free-energy landscape, principal component analysis, and structural clustering to identify representative conformations from molecular dynamics simulations: the myoglobin case. J. Mol. Graph. Model. **27**(8), 889–899 (2009)
23. Phillips, J.C., et al.: Scalable molecular dynamics with NAMD (2005). http://dx.doi.org/10.1002/jcc.20289
24. Pichaud, F., Walther, R.F., Nunes de Almeida, F.: Regulation of CDC42 and its effectors in epithelial morphogenesis (2019). https://doi.org/10.1242/jcs.217869
25. Plattner, N., Doerr, S., De Fabritiis, G., Noé, F.: Complete protein-protein association kinetics in atomic detail revealed by molecular dynamics simulations and markov modelling. Nat. Chem. **9**(10), 1005–1011 (2017)
26. Stelzl, L.S., Hummer, G.: Kinetics from replica exchange molecular dynamics simulations (2017). https://doi.org/10.1021/acs.jctc.7b00372
27. Stengel, K., Zheng, Y.: CDC42 in oncogenic transformation, invasion, and tumorigenesis (2011). https://doi.org/10.1016/j.cellsig.2011.04.001
28. Ung, P.M.U., Rahman, R., Schlessinger, A.: Redefining the protein kinase conformational space with machine learning. Cell Chem. Biol. **25**(7), 916–924 (2018)
29. Wang, J., Wolf, R.M., Caldwell, J.W., Kollman, P.A., Case, D.A.: Development and testing of a general amber force field. J. Comput. Chem. **25**(9), 1157–1174 (2004)
30. Wilson, K.F., Wu, W.J., Cerione, R.A.: Cdc42 stimulates RNA splicing via the s6 kinase and a novel s6 kinase target, the nuclear cap-binding complex (2000). https://doi.org/10.1074/jbc.c000482200
31. Wolf, S., Stock, G.: Targeted molecular dynamics calculations of free energy profiles using a nonequilibrium friction correction (2018). https://doi.org/10.1021/acs.jctc.8b00835

Using Molecular Dynamics to Assess How Two Insertion Mutations Affect Protein Structure

Changrui Li[1,2], Katie Christensen[1], Sarah Coffland[1], and Filip Jagodzinski[1(✉)]

[1] Western Washington University, Bellingham, USA
filip.jagodzinski@wwu.edu
[2] Old Dominion University, Norfolk, USA

Abstract. The functional impacts of mutations that result in one or more amino acid insertions into a protein have been the focus of recent studies. However, generating these mutations in a wet lab setting is time and resource intensive, and even prohibitive. Computational methods seek to bridge this gap by generating mutant proteins in silico. In this work we explore the use of molecular dynamics (MD) to assess the effects of pairs of insertion mutations into the PDB structure file of HIV-1 Protease. We use our in-house compute pipeline to generate the exhaustive set of mutants with two insertion mutations, and identify 24 mutant structures which present as outliers based on four structural metrics as reported in our earlier work. We present an analysis of the MD runs to show the extent that they reveal the effects of the insertion mutations that earlier work was unable to elucidate.

Keywords: Computational Proteomics · Applied Computing · Molecular Dynamics

1 Introduction

Double amino acid insertions or deletions (InDels) may result in complex and far-reaching effects on both the structure and function of a protein [17]. The need to better understand how pairs of InDel mutations affect a protein structure include studies to elucidate the functional and structural aspects of diseases caused by multiple mutations [3]. Due to the time and cost limitations of performing two insertion mutations in a physical protein, let alone creating the exhaustive set of all possible mutants with 2 insertion mutations in a wet lab setting, computational methods are often employed. These computational methods allow conducting large-scale and comprehensive surveys of the impacts of double mutations.

Computational approaches for generating protein mutants are inherently limited by the methodologies available to analyze them. One such method used to explore the behavior of *in silico* generated proteins is Molecular Dynamics (MD), which uses Newtonian equations of motion to simulate the behavior of systems with up to millions of particles [9]. One such popular program is GROMACS [2]. By running an MD simulation of a mutant generated by the software Rosetta [10], we are able to study the effects of the insertion mutations.

2 Methods

For this work, we used MD to study mutants with two insertion mutations into 99-residue aspartyl protease from the HIV-1 isolate BRU (PDB 1hhp). The exhaustive set of mutants for 1hhp was generated using a robotics inspired inverse kinematics approach [16] available in Rosetta.

2.1 Metrics

In our previous work [4,11,12], we demonstrated the use of four metrics (Table 1) – Rigidity Order Parameter (ROP) [1], Hydrogen Bond Count, Cluster Configuration Entropy (CCE) [14,15], and total energy as computed by Rosetta – to select outliers among the exhaustive set of mutants as a general approach to identify the most impactful insertion mutations. In Rosetta, the total energy unit is "Rosetta Energy Units" (REU), which is an arbitrary scale not directly comparable to physical energy units like kCal/mol, because it incorporates both physics- and statistics-based potentials. Both REU and kCal/mol however share the principle that a lower value specifies a more stable structure, with negative scores specifying the most stable conformations which are dominated by attractive, instead of repulsive, forces. ROP and CCE are calculated from the rigid clusters in a protein, which we identified using the KINARI-Lib software [7]. Rigidity analysis [8] identifies rigid clusters of atoms using an efficient combinatorial pebble-game approach.

Table 1. Metrics used to quantify impactful insertion mutations.

Metric	Explanation
Cluster Configuration Entropy (CCE)	A measure of the degree of disorder based on size of the largest rigid cluster [14].
Rigidity Order Parameter (ROP)	Protein flexibility calculated from the distribution of rigid clusters [1].
Hbond counts (HBC)	The total number of hydrogen bonds in the full protein as calculated by Rosetta [10].
Total Energy (REU)	Rosetta's energy term, physics- and statistics-based [10].

2.2 Mutants

To demonstrate the use of MD as a tool to gain insights about the effects of pairs of insertion mutations which our previous four metrics alone did not reveal [11,

12], we chose 24 mutants (Table 2) from among 2,020,000 in the exhaustive set of computer generated variants of 1hhp at three levels of impact. We relied on our compute pipeline which uses an inverse kinematics inspired approach and Rosetta to generate mutants with two insertions [5,16]. The three levels of impact include high impact (mutants with metric values falling in the 1st or 99th percentiles of all mutants), medium impact (mutants with metric values falling in the 75th or 25th percentiles), and low impact (mutants with metric values falling within the 50th percentile). High impact mutants are thus those whose metric values identify them as extreme outliers relative to the average metric value among all mutants, while low impact mutants are those whose metric values place them among the average metric values among all mutants. Throughout this manuscript, we designate mutants by specifying the two locations in the polypeptide chain where the insertions are made, and what amino acids are inserted. For example, 37P51M specifies inserting a Proline (P) between the existing residue 37 and 38, and a Methionine (M) between the existing residue 51 and 52, yielding a 101-residue mutant of 1hhp which normally has 99 residues. For the purpose of identifying mutant outliers based on hydrogen bonds, we made use of the count of hydrogen bonds as output by Rosetta when generating the mutants.

Table 2. Selected mutants using Cluster Configuration Entropy (CCE), Hydrogen Bond Counts (HBC) as determined using Rosetta when creating the mutants, Rigidity Order Parameter (ROP), and energy (Rosetta Energy Unit REU) metrics, their percentiles, and corresponding solvation (SOL) water molecules, where for example 32W96L designates a mutant in which a Tryptophan (W) was inserted at residue location 32, and Leucine (L) inserted at residue 96 of the wild type of 1hhp.

Metric	Val	%-ile	Mutation	SOL	Metric	Val	%-ile	Mutation	SOL
CCE	0.900	1	32W96L	9984	HBC	6.0	1	35V87F	9976
CCE	2.250	25	13W71D	9973	HBC	9.0	25	43K60L	12216
CCE	2.519	50	36I73H	9964	HBC	10.0	50	30D73Q	9959
CCE	2.519	50	8C48E	8691	HBC	10.0	50	37P51M	11741
CCE	2.622	75	62V92Q	9963	HBC	11.0	75	9A97A	9380
CCE	2.802	99	68C81Q	9978	HBC	14.0	99	2A99Y	9984
ROP	0.036	1	39V48W	13745	REU	−0.667	1	1R7I	13107
ROP	0.050	25	32I52C	9966	REU	−0.498	25	1T12N	9973
ROP	0.059	50	41L93A	13065	REU	−0.265	50	2H78W	10436
ROP	0.059	50	42F65H	12545	REU	−0.266	50	2P75H	9967
ROP	0.085	75	2C48G	10776	REU	0.183	75	14E61F	11304
ROP	0.213	99	6T58M	8624	REU	2.809	99	26Q59T	9957

2.3 MD Setup and Simulation Settings

For MD, we made use of GROMACS [2]. An explicit water solvent was used, with between 8,624 and 13,745 water molecules added to each simulation setup (Table 2, columns SOL). 50,000 steps where moles, volume and temperature (nvt) were kept constants, followed by 100,000 steps where moles, pressure, and temperatures (npt) were kept constant, were performed to equilibrate the system. This was followed by energy minimization for 1,000 steps. The MD run for each mutant was 10 million steps; each step was 2fs. Each MD run thus generated a 20ns simulation. Snapshots were saved every 500 steps of the MD run, during which the system's energy, count and location of hydrogen bonds, as well as the mutant's PDB structure, were saved.

2.4 Compute Resources

Each MD run took an average of 2 h and 29 min on a cluster configured with 8 compute nodes. Each node is equipped with dual Intel Xeon Gold 6130 CPUs, providing 16 cores and 32 threads per CPU (totaling 32 cores and 64 threads per node), and 192GB of DDR4 2666MHz ECC Registered RAM. The cluster also utilizes four Nvidia GeForce RTX 2080 TI GPUs (11GB GDDR6 memory, 4,352 CUDA cores) for GPU-accelerated processing, optimizing computation time. The nodes are connected through a 100Gb/s InfiniBand network for high-speed data transfer, ensuring efficient communication between nodes during parallel execution.

2.5 Energy Analysis of Mutants

We produce potential energy plots for all six mutants across all four metrics for all snapshots, calculated by GROMACS in kJ/mol.

2.6 Residue Centroid Analysis and Visualization

We average the (x, y, z) coordinates of the atoms that make up each residue, irrespective of their size or other biophysical properties, to compute a single residue "centroid" for each residue for every snapshot. For each residue, we then average the "centroids" across all snapshots, and calculate the central, or average, location of each residue throughout the simulation. Finally, we compute the distance between the residue's average (x, y, z) coordinates across all snapshots, and the residue's "centroid" (x, y, z) coordinates at each snapshot. To summarize the distributions of these distances across all residues, we produce Box and Whisker plots for each mutant protein as a way to infer the range of movement of every residue relative to its average "centroid" location. We make use of PyMol [6] to visualize the three-dimensional structures of a subset of the mutant proteins.

2.7 Hydrogen Bond Count Distribution Across Percentiles

As hydrogen bonds are essential to form secondary structures such as α-helices and β-sheets, which provide a protein with structural integrity, we selected to study them as indirect indicators of the effects of double insertion mutations. Because the count of hydrogen bonds varies over the course of an MD simulation, we calculated the count and locations of hydrogen bonds at each snapshot for all simulations using HBPLUS [13]. We group all Hydrogen Bond Counts at each percentile (1st, 25th, 50th, 50th, 75th, and 99th) for all four metrics (Cluster Configuration Entropy, Hydrogen Bond Count, Rigidity Order Parameter, and Rosetta Total Energy) from the MD runs. Thus, at each of the percentiles, we uncover how the Hydrogen Bond Count is distributed over all the snapshots as well as the outliers.

2.8 Observation of Hydrogen Bond Count Over Time

We create plots showing the variation of Hydrogen Bond Counts at each snapshot throughout the 10 million simulation steps, to determine how the Hydrogen Bond Count changed over time at each percentile.

3 Results and Discussion

As shown in Fig. 1, no meaningful correlation between the metric percentile of the mutant and its potential energy across snapshots is found. However, we note a large gap between the eventual lowest potential energy values of each mutant. For example, the mutants chosen by Rigidity Order Parameter show a wide range of asymptotic declines to a final total energy, with the mutant at the 0.01 percentile settling out to a total energy of less than $-700,000$ (kJ/mol), while the mutant at the 0.99 percentile does not even reach $-500,000$ (kJ/mol). These ranges in the total energies throughout the MD simulations suggest that using outliers alone for the different percentiles of the four metrics to identify most impactful pairs of insertion mutations does not correlate strongly with computed energy.

We plot the distributions of distances between each residue's average (x, y, z) coordinates across all snapshots, and the residue's "centroid" (x, y, z) coordinates at each snapshot as Box and Whisker plots (Fig. 2). We find three mutants that contain a subset of residues whose distances are greater than 60Å, specifying residues that moved significantly during the course of the simulation (Figs. 2a, 2c, and 2e). These mutants are 1R7I (top), 26Q59T (middle), and 39V48W (bottom). The three-dimensional structures of these mutants in Figs. 2b, 2d, and 2f, highlight the inserted residues (magenta) and the two residues (orange) which

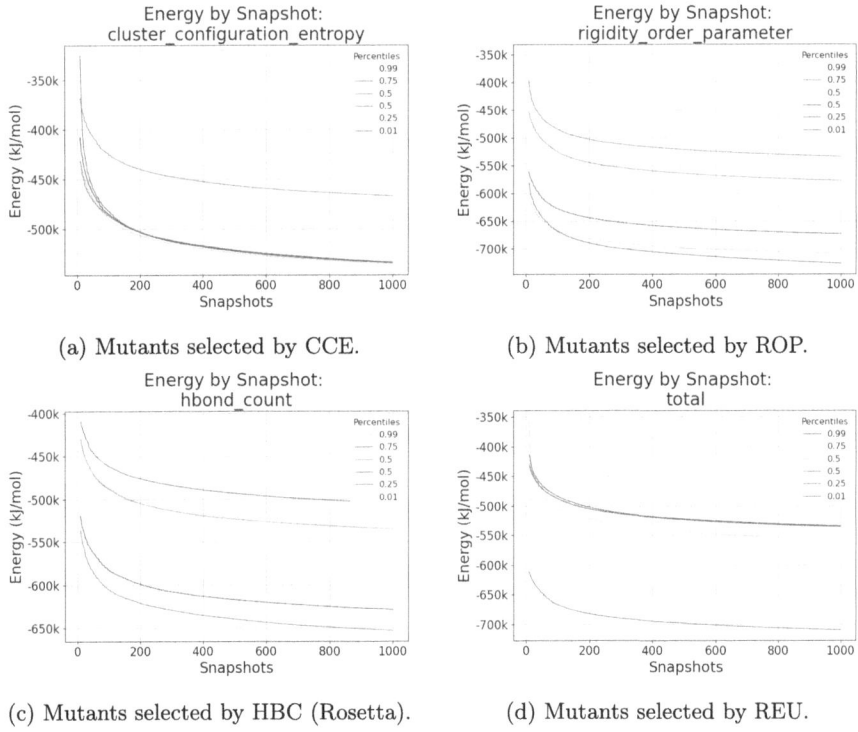

Fig. 1. Potential energy across snapshots (every 500-time steps) for each percentile for our four metrics.

present a whisker plot with the greatest number of outlier configurations relative to the average "centroid". The plots for all remaining mutants that we studied are in the Appendix (Sect. 6); the mutants there have "centroids" that presented motions less than 60Å (Fig. 5) from the average.

It is apparent that the insertion mutation's effects are not only local to the area where the insertions are made (orange), but in far away regions (orange) as well. For example, these trends are displayed in Fig. 2a. This mutant, 1R7I, was selected as a high outlier because it landed in the 1st percentile of the Rosetta energy relative to the energies of all other mutants generated. After the MD simulation, we found additional residues with high movement (additional outliers) demonstrating the high impact of its mutations across the protein.

In Fig. 3a, we explore how the number of hydrogen bonds changes as a function of MD simulation step at each of the individual percentiles. When a mutant is selected based on CCE, we find that at the 50th percentile, the mutant 36I73H has a higher Hydrogen Bond Count than all other selected mutants. The distribution at all percentiles have about the same range and the outliers are not far from the average. Figure 3b displays the Hydrogen Bond Count at the 99th percentile which has a wider distribution of outcomes and the outliers at this

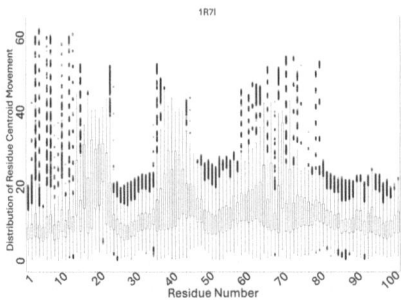

(a) Whisker plot for "centroid" for Mutant 1R7I, selected by Rosetta Energy, 1st %-ile.

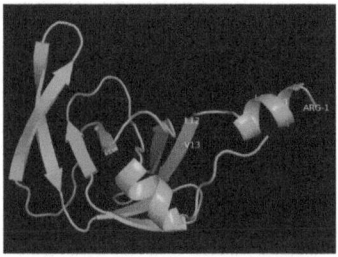

(b) Inserted residues (magenta) and most mobile residues (orange).

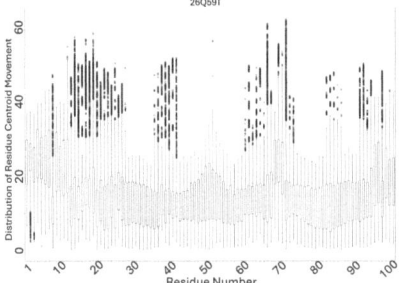

(c) Whisker plot for "centroid" for Mutant 26Q59T, selected by Rosetta Energy, 99th %-ile.

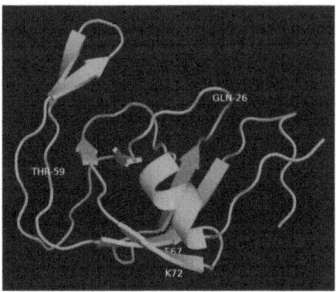

(d) Inserted residues (magenta) and most mobile residues (orange).

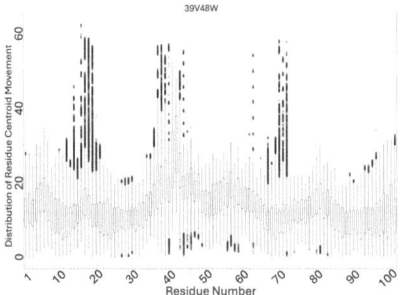

(e) Whisker plot for "centroid" for Mutant 39V48W, selected by ROP, 1st %-ile.

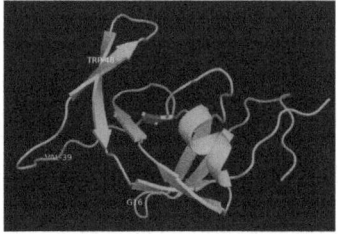

(f) inserted residues (magenta) and most mobile residues (orange)

Fig. 2. Distribution of movement as the distance between the residue's average (x, y, z) coordinates across all snapshots, and the residue's (x, y, z) coordinates at each snapshot (left). Three-dimensional structure of mutant proteins (right), highlighting the inserted residues (magenta) and the residues with high movement of more than 60Å (orange). (Color figure online)

level are even further from the mean. This indicates that the outliers experience an effect that distinguishes them from all other percentile levels. The opposite can be said for Figs. 3c and 3d.

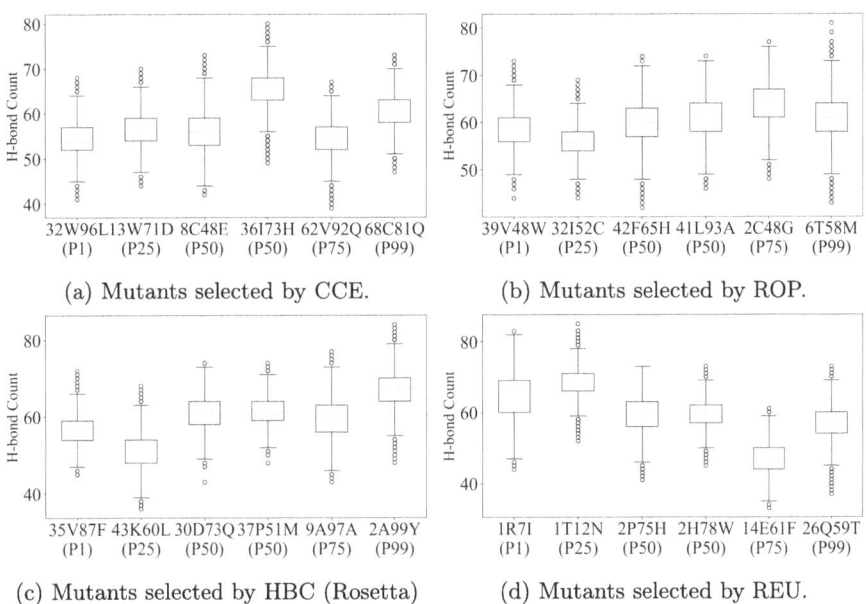

Fig. 3. Distribution of Hydrogen Bond Count of all snapshots at the 1st, 25th, 50th, 50th, 75th, and 99th percentiles. X-axis indicates the levels of percentiles in ascending order. Y-axis indicates the number of hydrogen bonds as calculated using HBPlus.

In Fig. 4, we see that the number of hydrogen bonds evolves over the course of the MD runs. Throughout most of the simulation, the Hydrogen Bond Count stays within the range of 50 to 70, with occasional spikes out of that range. In Fig. 4a, the mutant at the 50th percentile has a higher average value than other percentiles. In Fig. 4d, the mutants at the 99th and 25th percentiles have very different Hydrogen Bond Count averages as well. Since the mutants at the 99th and 1st percentiles are both high impact, this could indicate that these high impact mutations have features in them that cause Hydrogen Bond Counts outside the norm. In Fig. 4c, the Hydrogen Bond Counts start with a wider range and end up in a much tighter configuration. For Fig. 4b, all the lines are tangled together.

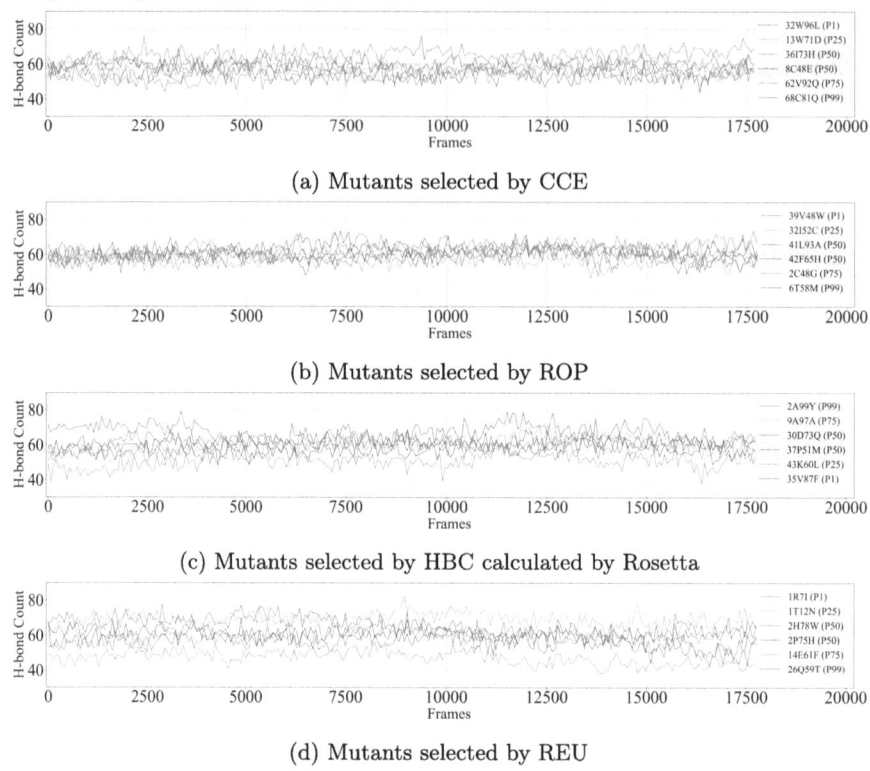

Fig. 4. Hydrogen Bond Count history. The x axis is the frames from 0 to 20,000 and y axis is Hydrogen Bond Count. The lines indicate Hydrogen Bond Count changes for different percentiles. Y-axis indicates the number of hydrogen bonds as calculated using HBPlus.

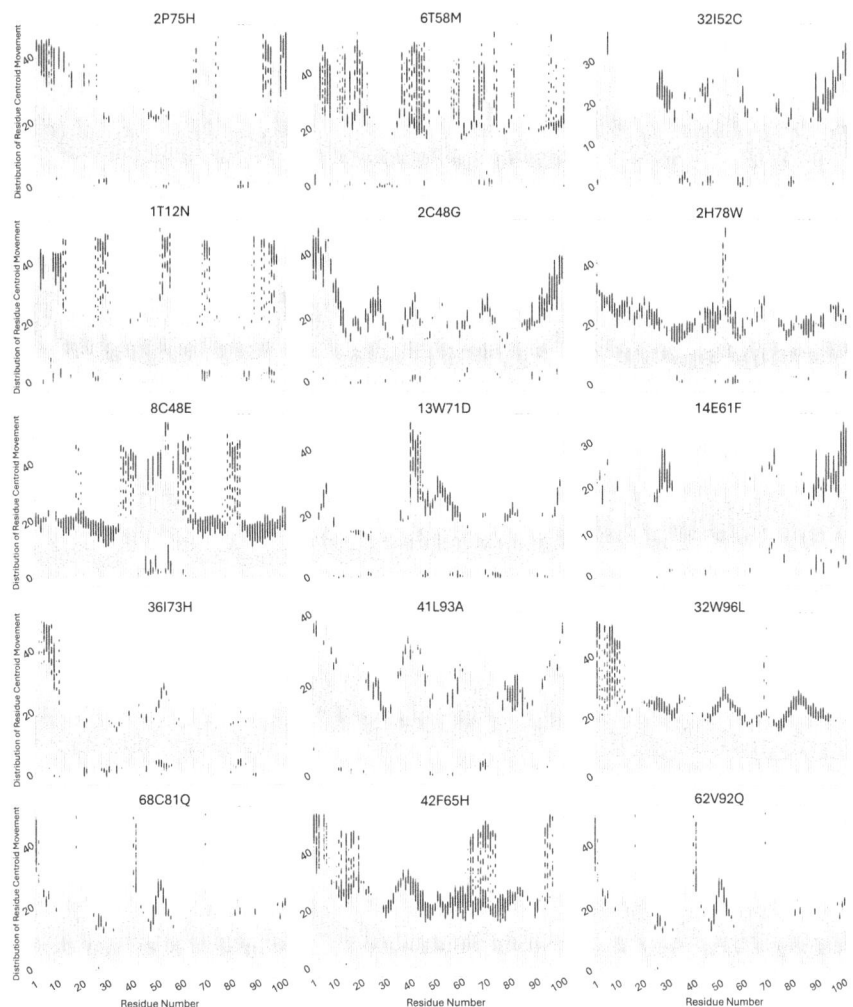

Fig. 5. Distribution of movement as the distance between the residue's average (x, y, z) coordinates across all snapshots, and the residue's (x, y, z) coordinates at each snapshot.

4 Conclusions

We show that selecting mutant proteins with double insertion mutations on a scale of mutation impact and simulating them using MD effectively reveals long- and short-range effects of the insertion mutations that other metrics alone don't reveal. Our simulations are for mutants with three levels of impact (high, medium, and low) selected using four metrics (Rigidity Order Parameter, Cluster Configuration Entropy, Hydrogen Bond Count, and Energy as calculated using Rosetta). We explore the relationships between influential mutations and the resulting energy and hydrogen bonds of the mutant proteins. We also investigate the comparative importance of utilizing one metric over another to determine mutants with high impact, and find that the metrics which correlate well with mutants in the 1st and 99th percentiles are the Rigidity Order Parameter metric and Rosetta energy score. The high impact mutants also display significant movement between the residue's average coordinates across all snapshots, and the residue's coordinates at each snapshot compared to the low and medium impact mutant proteins. We further demonstrate that the Hydrogen Bond Counts do not appear to have noticeable correlation with impact levels of the insertion mutations, though the potential energy of these mutants settle to drastically different levels.

5 Future Work

We recognize that these results do not correlate computational results with experimental results using methods such as deep alanine scanning, which we leave for future work. In this survey we didn't take into account hydrophobicity of amino acids, nor whether or not they are charged, which we leave to be explored in subsequent studies. Future computational experimentation includes increasing the number of energy minimization steps in order to settle the proteins more fully into a more stable energy before beginning the MD simulations. Additional future work includes a statistical investigation of the relationship between the levels of mutation impact, and the amount of residue movement within the mutant proteins throughout the MD simulations.

Additionally, scaling these experiments to larger proteins motivates future work in improving computational efficiency, such as applying deep learning to predict structural metrics like rigidity order parameters, or hydrogen bond counts, based on learned patterns in smaller protein structures. For this work, we comprehensively explored the double insertion mutation space for the 99 residue protein 1hhp, which comprised of 2,020,000 possible mutants. To exhaustively model a protein just 25 residues larger, we would need to generate more than 3,150,000 mutants, an increase of over 50%. The original work that generated these mutants for 1hhp took several weeks using a cluster of GPUs, further reinforcing the need for computational efficiency improvements to support scaling in future work.

Disclosure of Interests. The authors have no competing interests to declare that are relevant to the content of this article.

6 Appendix

The Box and Whisker plots displaying the distribution of residue movement of the mutant proteins excluded from Fig. 2 are below. These mutants do not contain outliers, therefore the distance between the residue's average (x, y, z) coordinates across all snapshots and the residue's (x, y, z) coordinates at each snapshot is less than 60Å.

References

1. Andersson, E., Hsieh, R., Szeto, H., Farhoodi, R., Haspel, N., Jagodzinski, F.: Assessing how multiple mutations affect protein stability using rigid cluster size distributions. In: 2016 IEEE 6th International Conference on Computational Advances in Bio and Medical Sciences (ICCABS), pp. 1–6. IEEE (2016)
2. Berendsen, H.J., van der Spoel, D., van Drunen, R.: GROMACS: a message-passing parallel molecular dynamics implementation. Comput. Phys. Commun. **91**(1–3), 43–56 (1995)
3. Choi, Y., Sims, G.E., Murphy, S., Miller, J.R., Chan, A.P.: Predicting the functional effect of amino acid substitutions and indels (2012)
4. Coffland, S., Christensen, K., Jagodzinski, F., Hutchinson, B.: RoseNet: predicting energy metrics of double indel mutants using deep learning. In: Proceedings of the 14th ACM International Conference on Bioinformatics, Computational Biology, and Health Informatics, pp. 1–9 (2023)
5. Das, R., Baker, D.: Macromolecular modeling with Rosetta. Annu. Rev. Biochem. **77**(1), 363–382 (2008)
6. DeLano, W.L., et al.: PyMOL: an open-source molecular graphics tool. CCP4 Newsl. Protein Crystallogr **40**(1), 82–92 (2002)
7. Fox, N., Jagodzinski, F., Streinu, I.: KINARI-Lib: a C++ library for mechanical modeling and pebble game rigidity analysis. Minisymposium on Publicly Available Geometric/-Topological Software, pp. 29–32 (2012)
8. Jacobs, D.J., Rader, A.J., Kuhn, L.A., Thorpe, M.F.: Protein flexibility predictions using graph theory. Proteins: Struct., Funct., Bioinf. **44**(2), 150–165 (2001)
9. Karplus, M., McCammon, J.A.: Molecular dynamics simulations of biomolecules. Nat. Struct. Biol. **9**(9), 646–652 (2002)
10. Leaver-Fay, A., et al.: Rosetta3: an object-oriented software suite for the simulation and design of macromolecules. In: Methods in Enzymology, vol. 487, pp. 545–574. Elsevier (2011)
11. Li, C., Jagodzinski, F.: Identifying impactful pairs of insertion mutations in proteins. In: Proceedings of the 14th ACM International Conference on Bioinformatics, Computational Biology, and Health Informatics, pp. 1–8 (2023)
12. Li, C., Zheng, Y., Jagodzinski, F.: How pairs of insertion mutations impact protein structure: an exhaustive computational study. Bioinf. Adv. vbae138 (2024)
13. McDonald, I.K., Thornton, J.M.: Satisfying hydrogen bonding potential in proteins. J. Mol. Biol. **238**(5), 777–793 (1994)

14. Pfleger, C., Radestock, S., Schmidt, E., Gohlke, H.: Global and local indices for characterizing biomolecular flexibility and rigidity. J. Comput. Chem. **34**(3), 220–233 (2013)
15. Radestock, S., Gohlke, H.: Exploiting the link between protein rigidity and thermostability for data-driven protein engineering. Eng. Life Sci. **8**(5), 507–522 (2008)
16. Turcan, A., Chou, G., Martin, L., Miller, T., Thompson, D., Jagodzinski, F.: Exhaustive in-silico simulation of single amino acid insertion and deletion mutations. In: 2022 IEEE International Conference on Bioinformatics and Biomedicine (BIBM), pp. 3498–3503. IEEE (2022)
17. Wolf, Y., Madej, T., Babenko, V., Shoemaker, B., Panchenko, A.R.: Long-term trends in evolution of indels in protein sequences. BMC Evol. Biol. **7**, 1–10 (2007)

Toward Modeling Protein Multimers by Combining AlphaFold 3 Predictions with Secondary Structures from Medium-Resolution Cryo-EM Maps

Changrui Li[1], Thu Nguyen[1], Willy Wriggers[2], and Jing He[1(✉)]

[1] Department of Computer Science, Old Dominion University, Norfolk, VA 23529, USA
jhe@cs.odu.edu
[2] Department of Mechanical and Aerospace Engineering, Old Dominion University, Norfolk, VA 23529, USA

Abstract. AlphaFold 3 (AF3) has recently been shown to offer improved accuracy in predicting the structures of protein multimers. Improved models may lead to new opportunities for fitting them to cryo-electron microscopy (cryo-EM) maps with medium resolution (5–10 Å). Deriving atomic models from such cryo-EM maps is still challenging due to the lack of high-resolution features. Our case study involving four AF3 multimer models and corresponding cryo-EM maps with 7–8 Å resolution showed that the predicted multimer models were partially correct. The predicted models contained fairly accurate domains, secondary structures, and individual chains, since 9 of the 17 chains exhibit TM-scores higher than 0.8 and 16 chains had TM-scores above 0.5 compared with the official atomic structures that were deposited with the cryo-EM maps. However, some cases exhibited incorrect relative positions of individual chains or domains. We observed that the order of cross-correlation (CC) scores between the multimers and their corresponding cryo-EM maps aligned with the order of the TM-scores. This shows that if regions are masked correctly, CC scores are sensitive enough to distinguish among the multimer models. A masking of monomeric chains may not always be attainable, so we also explored the level of accuracy in secondary structure segmentation for one of the cases in greater detail. Although molecular details are not fully visible in cryo-EM maps at medium resolution, the location of major secondary structures, such as α-helices and β-sheets, were detectable using our DeepSSETracer tool. Our analysis illustrates the potential for improvements in the accuracy of AF3-predicted multimer models by combining the density map–model similarity (CC scores) and the secondary structure map–model similarity in a future approach.

Keywords: Protein Structure · Structure Validation · Fitting · Cryo-Electron Microscopy · AlphaFold · Secondary Structure

1 Introduction

Over the last decade, it has become routine practice to obtain atomic structures from cryo-electron microscopy (cryo-EM) maps at 4 Å resolution or better [1, 2]. However, it is still challenging to derive atomic structures accurately from medium-resolution (5–10 Å) cryo-EM maps. Due to the lower quality of such maps, the side chains of protein are not visible, and it is challenging to distinguish the backbone of a protein. Atomic structures derived from medium-resolution cryo-EM maps have primarily been obtained through rigid-body or flexible fitting of existing structures known to be structurally similar to the target protein [3, 4]. After fitting, some models are refined to optimize the agreement of local map density and derive chemical properties using tools such as Phenix [5].

Due to the poor quality of the medium-resolution maps, it is challenging to validate structures derived from such maps. There are quality measures for assessing the health of atomic models, such as clash and rotamer scores, but they are not linked to cryo-EM maps' characteristics, and therefore, it is not clear how accurate such models interpret cryo-EM maps. Our recent survey of local structure quality using the MolProbity score showed that atomic structures solved from medium-resolution cryo-EM maps had overall higher (worse) MolProbity scores than corresponding models in the AlphaFold (AF) Database [6]. The survey results suggest that the AF-predicted models have an overall more valid local structural configuration than Protein Data Bank (PDB) structures obtained through fitting and refinement using medium-resolution cryo-EM maps. The overall accuracy of PDB models still offers room for improvement when better modeling methods are used. Before AF became available, modeling methods generally relied on fitting or modifying known PDB templates expected to be similar to the unknown (target) protein structure. This is different from what AF provides, which is a model predicted directly from amino acid sequences of the target protein. With the availability of a better starting model, it is likely that more accurate models can be obtained by revisiting older medium-resolution cryo-EM maps and their initially deposited PDB models.

AF2 has been shown to be the best overall method of protein structure prediction at past Critical Assessment of Protein Structure Prediction (CASP) conferences [7, 8]. Models predicted using AF2 have also been used to derive atomic structures from cryo-EM maps [9, 10]. AF3, which was made available recently, specifically improved the accuracy of multimer predictions [11]. In the present study, we conducted case studies to understand the level of accuracy of predicted AF3 multimers and individual chains with respect to known (officially deposited) models and their corresponding medium-resolution cryo-EM maps. The similarities between the AF3-predicted models and their corresponding cryo-EM map regions were estimated using cross-correlation (CC) scores, whereas similarities with the PDB-deposited structures were assessed by Template Modeling scores (TM-scores). Insights from this study are expected to inform the design of a better modeling approach for medium-resolution cryo-EM maps in the future.

2 Methods

Four proteins with PDB IDs 4CG5, 5FUA, 5FWP, and 5VHW were used in the case studies below. The atomic structures of these four proteins were initially derived from cryo-EM maps with resolutions between 7 Å and 8 Å. The corresponding cryo-EM maps were downloaded from the Electron Microscopy Data Bank (EMDB) [12]. The amino acid sequences of all chains in each PDB structure were submitted to the AF3 (beta version) prediction tool to obtain an multimer model [11]. The predicted multimer model with the highest-ranking score was selected for further evaluations.

2.1 Evaluation of TM-Scores of the AF3-Predicted Multimers Using MM-Align

The free AF3 online service (beta version) was used to obtain the predicted models for each multimer [11]. For convenience, the AF3-predicted multimer models were converted from .cif format to .pdb format using Phenix [13]. MM-align is an algorithm for comparing the structural similarity of multimers [14]. We used MM-align to calculate the TM-score between an AF3-predicted multimer and the corresponding atomic structure in the PDB. A TM-score ranges between 0 and 1, with a score above 0.5 often suggesting that two models share the same fold.

2.2 Evaluation of Individual Chains of AF3-Predicted Multimers Using TM-Align

To examine the level of accuracy of individual chains in the predicted multimers, the atomic coordinates of each chain were extracted from the AF3 multimer model and used for calculation of a TM-score against its corresponding chain in the PDB structure. TM-align was used to calculate the TM-score for individual chains [15].

2.3 Evaluation of CC Between Predicted Multimers and Corresponding Cryo-EM Maps

CC scoring functions have routinely been used to measure the similarity between a model and a cryo-EM map. A model is first converted to a density map by simulating each atom with a point cloud of a certain radial distribution that corresponds to the desired experimental resolution level. Then the CC is computed between the simulated (lower-resolution model) map and the experimental map. We measured the CC score between an AF3 multimer model and the cryo-EM map that corresponds to the PDB structure used in each case. The *collage* utility of the Situs package was used to calculate the CC score at a given relative position of the atomic model relative to the map [16]. The specific alignment between a predicted AF3 multimer and the cryo-EM map was produced using MM-align [14]. Since the official PDB structure was derived from its cryo-EM map and provided a reliable frame of reference, the predicted model was aligned according to the known PDB structure.

To reduce the CC score's influence in the map regions unrelated to the PDB structure, the cryo-EM map was masked using the PDB structure as a reference using ChimeraX [17]. The mask radius was 5 Å around each atom. By doing so, the CC score reflected the similarity between the AF3-predicted model and the map region defined by the PDB structure. We also computed the CC score between a chain in an AF3-predicted multimer and its corresponding map region defined by the PDB chain. The extraction of the masked density region was performed using ChimeraX, which used the PDB chain structure as a reference.

2.4 Detection of Secondary Structure Elements from the Cryo-EM Map Corresponding to PDB ID 5VHW

DeepSSETracer was used to detect helices and β-sheets from the cryo-EM map for PDB ID 5VHW [18, 19]. As a preprocessing step for DeepSSETracer, the cryo-EM map of EMD-8685 was resampled to 1 Å in ChimeraX [17]. Since DeepSSETracer was designed for local regions of a cryo-EM map, the original size of 360 × 360 × 360 was first restricted to its molecular region of 108 × 155 × 179 in size. A preprocessing step was added to further split it into four sub-maps, each with a size allowed by the memory of the computer. DeepSSETracer was applied to each sub-map, and the segmentation results were averaged for the overlapping region.

3 Results

To understand the accuracy of predicted multimer models and the similarity between the models and corresponding cryo-EM maps, we investigated four cases that contained multiple chains. Two of the cases (5VHW and 5FUA) contained multiple copies of the same chain sequence, and the other two (4CG5 and 5FWP) contained a longer chain and two shorter chains in each. The investigation focused on four aspects: the accuracy of the individual chain models of the predicted multimers, the accuracy of predicted multimers, the CC score between a predicted multimer model and the cryo-EM maps, and secondary structure segmentation from the corresponding cryo-EM maps.

3.1 Individual Chain Models Predicted Using AF3

The predicted multimer models were examined at the individual chain level. The coordinates of atoms of each chain in a predicted multimer were extracted and compared with its corresponding chain in the PDB structure. For example, in the case of PDB ID 4CG5, chain A in the predicted multimer was aligned with chain A of the PDB structure using TM-align, and the TM-score was 0.75 when the PDB chain A was used as a reference (Table 1). The highest TM-scores were 0.93 and 0.96 for the two copies (A chain and C chain) of the four chains in PDB ID 5VHW (Table 1). In addition to computing numeric TM-scores, we also used the Matchmaker tool of ChimeraX to visualize the alignment. The predicted model (gold) for chain A aligned very well with its corresponding chain in the PDB structure (cyan), as suggested by the high TM-score of 0.93 (Fig. 1B). The best-aligned domains, shown in the side view, were the left and middle domains (dark blue arrows in Fig. 1B). The right domain aligned less well among the three, although the long helices were clearly predicted. The slight misalignment of the right domain was likely due to the loop linking the right and middle domains, which is a known challenge to predict accurately. In chain A of the predicted model, there was a portion of the right domain that was not included in the atomic structure of PDB ID 5VHW (red arrow in Fig. 1B). This segment was not resolved in the PDB structure, most likely because the cryo-EM map quality was poor in that region. (It is common to see variation in map quality across cryo-EM maps.) Since the TM-score was calculated using the PDB chain as a reference, the extra portion in the predicted chain did not affect the TM-score. In fact,

Table 1. The TM-scores and CC scores of AF3-predicted multimer models and individual chains. From left to right: the PDB ID, the number of amino acids in the chain, EMDB ID (EMD), resolution of the cryo-EM map, the TM-scores of the AF3 multimers compared to the PDB structures, the TM-scores of individual chains, the CC scores between the AF3 multimers and the cryo-EM maps, and the CC scores of individual chains in the AF3 multimer model.

PDB ID	Chain ID, Length	EMD	Res	MM-align (Multimer)	TM-align (Individual chain)	CC (Multimer)	CC (Individual chain)
4CG5	A:476B: 68C: 36	2511	7.4 Å	0.77	A: 0.75 B: 0.62 C: 0.38	0.72	0.70 0.67 0.55
5FUA	A, B, C, D, E, F: 362	3283	7.6 Å	0.34	A: 0.91 B: 0.82 C: 0.81 D: 0.81 E: 0.84 F: 0.81	0.39	0.76 0.70 0.70 0.70 0.69 0.70
5FWP	A, B: 727C: 378D: 310	3340	7.2 Å	0.64	A: 0.69 B: 0.69 C: 0.63 D: 0.83	0.71	0.73 0.73 0.66 0.84
5VHW	A, B, C, D: 1057	8685	7.8 Å	0.60	A: 0.93 B: 0.52 C: 0.96 D: 0.52	0.48	0.73 0.53 0.75 0.53

the TM-scores for both the A and C chains were 0.93 and 0.96, respectively, showing great similarity between the predicted and experimentally derived chains.

We examined the individual chains that had low TM-scores of 0.62, 0.38, 0.52, and 0.52, respectively. Both chains B and C of PDB ID 4CG5 were small chains containing one to two helices. Although the helices were predicted by the model, they were predicted at different orientations from those in the PDB structure. The two other low TM-scores belonged to the predicted B and D chains of PDB ID 5VHW. Although the predicted A chain had a high TM-score of 0.93 and aligned well with the A chain of PDB ID 5VHW (Fig. 1B), the B chain had a low TM-score of 0.52 and did not align well with chain B of PDB ID 5VHW (Fig. 1C). In fact, only the two left domains aligned well (dark blue arrows in Fig. 1C), but the alignment started to break at the loop linking the middle domain. Due to the flexibility of the loop, the middle and right domains had a different geometric arrangement than in the PDB structure, regardless of the high similarity within the domains. Note that the A and B chains in the PDB structure showed a different arrangement (cyan in Fig. 1B,C), even though they had the same amino acid sequence. In contrast, the predicted A and B chains had a similar domain arrangement (gold in Fig. 1B,C), which wrongly predicted a similarity between the A and B chains.

The TM-score was calculated for each pair of chains. Among the 17 chains in the four cases, 16 exhibited TM-scores above 0.5, and 10 were above 0.7. Generally, a TM-score above 0.7 suggests overall good similarity between two models, and a score above 0.5 suggests the same fold. Individual chains in the four cases were predicted well in relation to the AF3 multimers. Although there was inaccuracy in some loops, resulting in errors in the relative geometric arrangement of domains separated by those loops, such errors were likely to be repaired when cryo-EM maps were included in the refinement, since cryo-EM maps at medium resolution generally contain enough detail to distinguish domains and even major secondary structure elements.

3.2 Predicted Multimer Models Using AF3

The multimer models were compared to their corresponding atomic structures in the PDB using the MM-align tool [14]. MM-align was designed to measure the overall TM-score for protein complexes with multiple chains. It is a method extended from TM-align that compares two individual chains. We observed much lower TM-scores for the full multimers than for some of the individual chains. These TM-scores were 0.34, 0.60, 0.64, and 0.77, respectively, for the four cases (Table 1). The lower TM-scores of the multimers reflects the challenges in predicting complex multimer structures with AF3, since such predictions require additional information (more than just the amino acid sequence) to correctly place multiple chains and domains relative to each other.

The highest TM-score of 0.77 belonged to the case of PDB ID 4CG5, which is dominated by a long A chain. The shorter B chain of this case contains two helices, and the shortest C chain contains only one helix. The second-highest overall TM-score, 0.64, was observed for PDB ID 5FWP, which contains two copies of a longer chain (727 residues) and two shorter chains (378 and 310 residues, respectively). Figure 1A shows the alignment between the AF3-predicted multimer (gold) and the atomic structure (cyan). In this case, two (A and C chains) of the four copies aligned much better with their corresponding chains in the atomic structure, but chains B and D did not, as can be observed in Fig. 1A, where the gold ribbons B' and D' are exposed and lie outside the cyan ribbons. The TM-score for the AF3 multimer was only 0.60 for the full PDB ID 5VHW, suggesting significant disagreement when all four chains are considered collectively.

3.3 Inaccuracies in Predicted AF3 Multimers

The TM-scores measured at the chain level were substantially higher than those measured at the multimer level for two of the four cases. We examined these two cases to understand the underlying reasons for this discrepancy. As mentioned in Sect. 3.2 for PDB ID 5VHW, when the predicted multimer was aligned with the PDB structure, chains A and C were well-aligned with their corresponding chains in the PDB structure (Fig. 1D). However, the top view of the alignment shows that the predicted chain B (marked as B' in Fig. 1E) was offset clockwise in the figure. It appears that the four chains in the PDB structure were more tightly associated than the predicted multimer. It is likely that the incorrect domain arrangement of chains B and D (Sect. 3.1) contributed to the overall inaccuracy

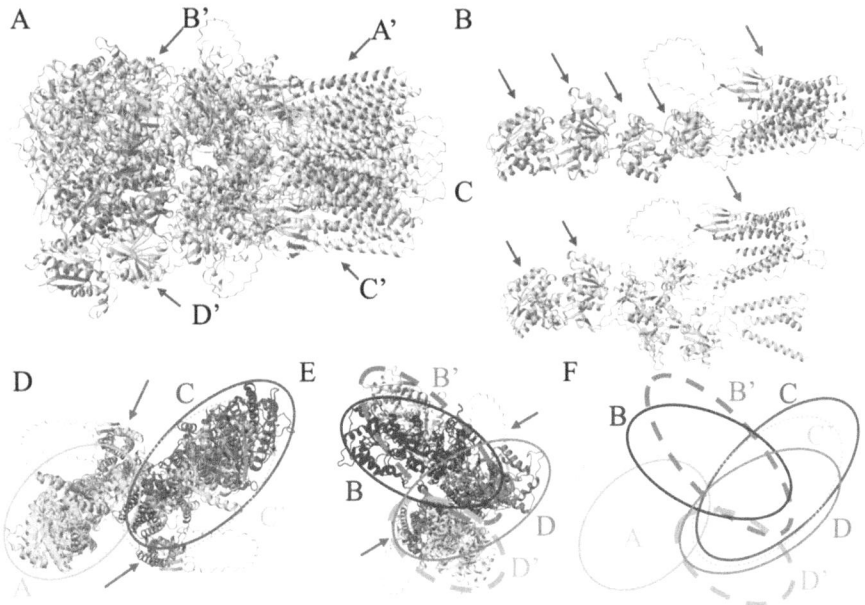

Fig. 1. Atomic structure of PDB ID 5VHW and the corresponding multimer model predicted by AF3. (A) Side view of PDB structure 5VHW (cyan) aligned with the predicted multimer model (gold) using the MM-align tool. The regions of the predicted model (chains A' to D') that do not align with the PDB structure (chains A to D) are indicated with arrows and labels. (B) Side view of the superposition of chain A in PDB ID 5VHW with chain A in the predicted model. (C) Side view of the superposition of chain B in PDB ID 5VHW with chain B in the predicted model, annotated similarly as in (B). (D) Top view of chains A (orange) and C (magenta) in PDB ID 5VHW aligned with the predicted chains A and C, denoted as A' and C', respectively. (E) Top view of chains B (blue) and D (green) in PDB ID 5VHW aligned with the predicted chains B and D, denoted as B' and D', respectively. (F) Positions of chains A (orange), B (blue), C (magenta), and D (green) in PDB ID 5VHW (solid ovals) and the predicted chains (dotted ovals), as seen from the top. In Panels D, E, and F, the chain structures of PDB ID 5VHW are displayed in a darker shade, while those in the predicted model are colored in dotted/dashed lines. The dark blue arrows in (B-E) indicate well-aligned domains. The red arrows indicate structural segments present only in the predicted models and not in the PDB structure. (Colour figure online)

in chain positions. This is an example where the intrinsic flexibility of a biological system limits the accuracy that can be achieved by sequence-based protein structure prediction.

The lowest TM-score measured at the multimer level was 0.34 for PDB ID 5FUA (Table 1). In this case, the individual TM-scores were between 0.81 and 0.91 for the six chains, suggesting good similarity at the local chain level. However, the arrangements of the six copies of the same chain sequence were quite different (Fig. 2A,B). It appears that the PDB structure had a clear five-fold symmetry for chain 1 through chain 5 surrounding a central cavity (Fig. 2A). However, there was no five-fold symmetry in the predicted multimer (Fig. 2B). Therefore, the TM-score for the predicted multimer was low when it was directly compared with the PDB structure. However, an examination together with the cryo-EM map showed that the PDB structure only provided the six chains in an asymmetric unit, and there were many more chains of the structure related by a specific symmetry. The PDB structure was focused on one ring (R_1 in Fig. 2C), but the predicted multimer had the six chains in three rings (R_1, R_2, and R_3 in Fig. 2C). It is possible that the predicted multimer would have had a higher TM-score if it were compared with a different structure where chains are remapped to the corresponding locations, but our study was limited to the measuring of similarity against the PDB structure. This case demonstrates a challenge in protein complex structure prediction when there are symmetrically related chains involved.

3.4 Similarity Between the Predicted Models and Cryo-EM Maps

To improve a predicted model using a cryo-EM map, the predicted model first needs to be registered in the map, and CC is a commonly used measure to select the optimal registration. To get an idea of the level of agreement between a predicted multimer and its corresponding cryo-EM map, we calculated the CC score between them. A predicted multimer model was first aligned with the PDB structure using MM-align, and then the CC score was calculated using the *collage* utility of Situs [16]. As an example, the AF3-predicted multimer was first aligned with the PDB structure 5VHW so that it is best aligned with the cryo-EM map. The CC score in this case was 0.48, which reflects the fact that only two of the four predicted chains aligned well with the PDB structure (see Sect. 3.2). In addition, the cryo-EM map had a poor-quality region near the right domain of PDB ID 5VHW (red arrow in Fig. 3A). Note that the CC scores for the two better-aligned chains A and C were 0.73 and 0.75, which is higher than those of the badly aligned chains B and D (0.53 and 0.53, respectively). Unlike calculating the CC score between a predicted multimer and the cryo-EM map, the calculation of the CC score of an individual chain requires the map region of the individual chain to be masked out from the entire map using the PDB chain as a reference. Therefore, the CC score of an individual chain shows how similar a predicted chain is with the map region carved out using existing knowledge about a PDB chain. The difference between the two CC scores (0.73 and 0.53) suggests that if a map region of a chain is correctly identified, then the CC score is sensitive enough to distinguish between the chain conformations of the A and B chains, in which the domain arrangements are different (Fig. 1B,C).

In the case of PDB ID 5FUA, the CC score between the predicted multimer and the cryo-EM map was 0.39. The low score reflects the completely different overall position of the six chains in the predicted multimer compared with the PDB structure (see Sect. 3.3). The CC scores for the six individual chains were much higher, between 0.69 and 0.76, suggesting that they are much more similar to the map regions individually. Overall, the CC scores for the complete multimers in the four cases were 0.39, 0.48, 0.71, and 0.72.

3.5 Potential of Utilizing Medium-Resolution Cryo-EM Maps for Refining AF3-Predicted Multimers

We observed that AF3-predicted multimers may contain errors in the global arrangement of individual chains or domains, but the local structure of these chains or domains was more accurately predicted. To use medium-resolution cryo-EM maps to refine predicted multimers, a suitable selection method is needed to identify the best global arrangement

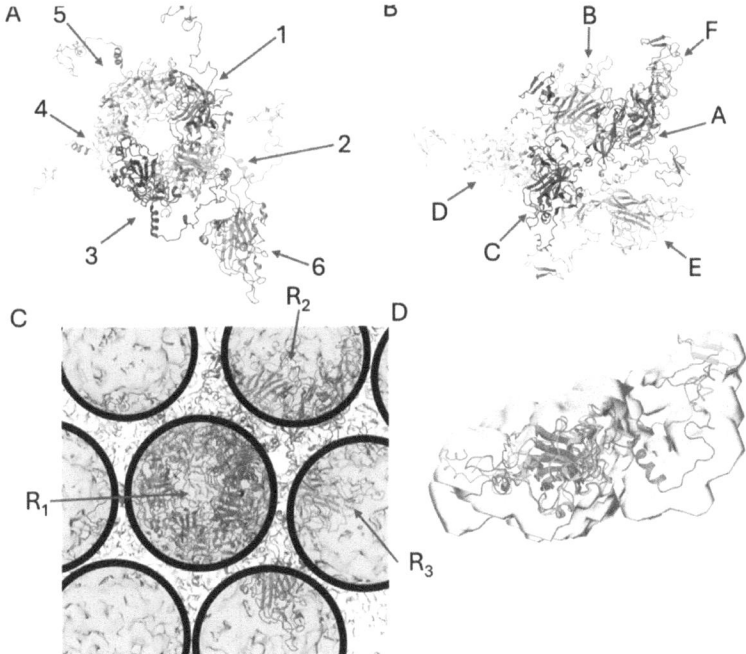

Fig. 2. The atomic structure PDB ID 5FUA and the AF3-predicted multimer model. (A) the six chains (1–6) in the PDB structure 5FUA; (B) AF3-predicted multimer structure that includes chains A to F; (C) The PDB structure (in red) was aligned with the AF3-predicted multimer (six colored ribbons) using MM-align and was superimposed on the cryo-EM map (EMD-3283, transparent gray volume). Rings (solid black lines) were drawn to indicate the five-fold symmetrical copies of the same chain sequence. The six chains in the predicted multimer are spread across three ring regions: R_1, R_2, and R_3. (D) Chain B (green) of the predicted multimer was aligned using Matchmaker in ChimeraX with its corresponding chain in the PDB structure (red) and was superimposed with the masked region of the cryo-EM map corresponding to the PDB chain. (Colour figure online)

and distinguish it from alternatives that require only minor local translations or rotations of the monomeric fragments. It is unclear as to whether our current measure using CC scores is sensitive enough to serve this purpose. Our study only measured a near-ideal situation, when the correct outline of map regions was provided in the calculation of CC scores. In future work, we plan to explore if CC scores can distinguish among different models involving slightly different arrangements. Nevertheless, it was more straightforward in the present work to take advantage of additional secondary structure propensities, in addition to mass densities measured by the CC score.

Since secondary structure elements, particularly long helices and larger β-sheets, are computationally recognizable in medium-resolution cryo-EM maps, we explored their suitability for characterizing the arrangement of individual chains in one of the four cases, PDB ID 5VHW. The cryo-EM map EMD-8685 (transparent gray) is shown in Fig. 3A. Using DeepSSETracer, most of the helix regions were correctly detected (yellow in Fig. 3B). The major locations of β-sheets were also detected. We noticed that

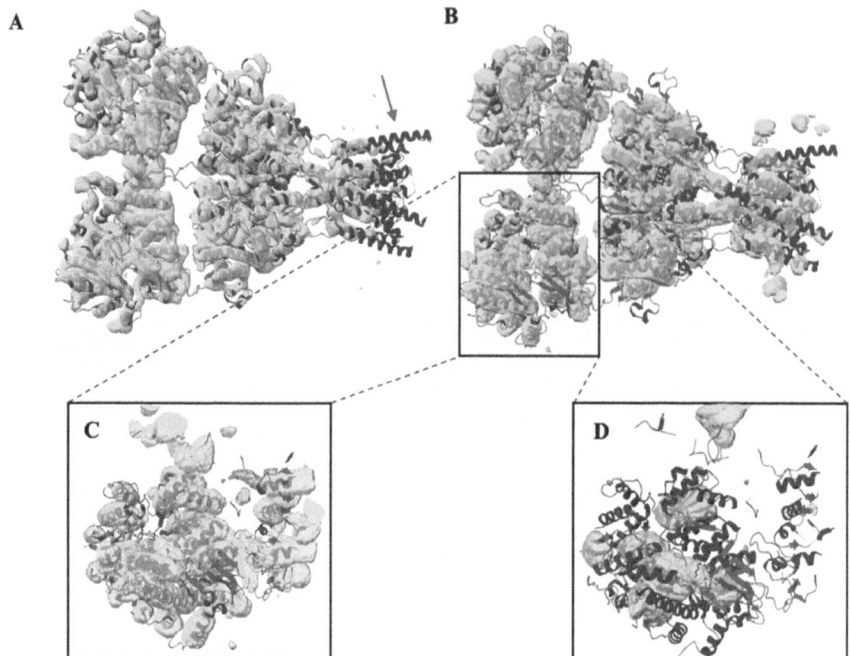

Fig. 3. Secondary structure segmentation of EMD 8685 using DeepSSETracer [19]. (A) Cryo-EM map EMD 8685 and its corresponding structure PDB ID 5VHW (helices: blue; strands: magenta; coil: gray) superimposed in ChimeraX [17]. The red arrow in (A) indicates a weak density region of the map. (B) Segmentation result from DeepSSETracer for helices (yellow) and β-sheets (cyan) superimposed with the PDB structure. (C) The enlarged view of a region showing detected helix regions (yellow). (D) The enlarged view of the same region showing detected β-sheet regions (cyan). (Colour figure online)

the map quality was better in the middle and at the left side of the map in Fig. 3, but the right side had much weaker density (arrow in Fig. 3A). The resulting helix segmentation using DeepSSETracer also missed the secondary structure density around that location (Fig. 3B). In an enlarged view (Fig. 3C), it can be observed that the detection of helices was mostly correct, while β-sheet detection (Fig. 3D) as attenuated compared to the PDB structure. Nonetheless, DeepSSETracer shows that this cryo-EM map contained sufficient detail for assigning secondary structures. It is therefore reasonable to expect that the matching of both densities and secondary structure propensities in a future "meta-CC" approach will provide a discriminative measure that facilitates the correct placement of AF3-predicted chains.

4 Conclusion

Deriving atomic structures from cryo-EM maps at medium resolution is still a challenging problem due to the relatively low quality of such maps. However, AF3-predicted models and the cryo-EM maps used complementary information about protein structures that could be combined for better modeling outcomes. Cryo-EM maps, even at the medium resolution, contain the global positions of locally dense regions that correspond to secondary structure elements. The AF3-predicted models contained fairly accurate local models, in terms of chains, domains, and secondary structures. How to utilize both types of information to improve the accuracy of modeling is an interesting and challenging problem.

The quality of structures derived from the medium-resolution cryo-EM maps is often limited by the quality of the cryo-EM map itself and the quality of the template structure used for fitting the map. With the advancement of protein structure prediction using AF3, it is important to understand its potential impact for cryo-EM maps of medium resolution. In our investigation of four cases using cryo-EM maps between 7 Å and 8 Å in resolution, we observed that the AF3-predicted multimer structures were partially correct, showing TM-scores of 0.34, 0.60, 0.64, and 0.77. Additionally, 16 of the 17 chains in the four cases showed TM-scores above 0.5, suggesting that the folds were predicted correctly. Some of the individual chains even exhibited TM-scores above 0.8 compared with the corresponding chains in the PDB structure, suggesting that they can be used directly with minimal changes when combined with cryo-EM maps. However, the global arrangement of the monomeric fragments in some AF3-predicted multimer models requires further refinement.

In addition to the investigation of the accuracy of AF3-predicted multimers and individual chains, we explored two aspects in utilizing predicted models in cryo-EM maps at the medium resolution. We observed that if the correct outline of a chain in a cryo-EM map was provided, the CC scores were a good predictor of accuracy (since they showed the same order as TM-scores at the individual chain level). This observation suggests that if regions are masked correctly, CC scores become sensitive enough to distinguish among multimers. Therefore, in a near-ideal situation, CC scores would allow us to predict correct model placement. However, we did not test if CC scores are sensitive enough when the outline of a chain is unknown, as is often the case at medium resolution. In practice, we likely need a more sensitive measure to distinguish two models in which the domains or chains exhibit a different arrangement. We showed evidence that some major secondary structures can be detected well in medium-resolution maps. Our future goal is to include secondary structures detected from cryo-EM maps to design more sensitive selection methods for AF3 multimer prediction and refinement. The analyses conducted in relation to the four case studies above will help inform the better design of future approaches to obtain atomic structures from medium-resolution cryo-EM maps.

Funding. This work was supported by National Institutes of Health Grants R01-GM062968 and R35-GM153431 and the Old Dominion University Batten Endowment to W.W.

Disclosure of Interests. The authors have no competing interests relevant to the contents of this article to declare.

References

1. Zhang, K., Pintilie, G.D., Li, S., Schmid, M.F., Chiu, W., Zhang, K., et al.: Resolving individual atoms of protein complex by cryo-electron microscopy. Cell Res. **30**(12) (2020). https://doi.org/10.1038/s41422-020-00432-2
2. Fromm, S.A., O'Connor, K.M., Purdy, M., Bhatt, P.R., Loughran, G., Atkins, J.F., et al.: The translating bacterial ribosome at 1.55 Å resolution generated by cryo-EM imaging services. Nat. Commun. **14**(1) (2023). https://doi.org/10.1038/s41467-023-36742-3
3. Trabuco, L.G., Villa, E., Mitra, K., Frank, J., Schulten, K.: Flexible fitting of atomic structures into electron microscopy maps using molecular dynamics. Structure. **16**(5), 673–683 (2008). https://doi.org/10.1016/j.str.2008.03.005
4. Kovacs, J.A., Galkin, V.E., Wriggers, W.: Accurate flexible refinement of atomic models against medium-resolution cryo-EM maps using damped dynamics. BMC Struct. Biol. **18**(1), 12 (2018). https://doi.org/10.1186/s12900-018-0089-0
5. Afonine, P.V., Poon, B.K., Read, R.J., Sobolev, O.V., Terwilliger, T.C., Urzhumtsev, A., et al.: Real-space refinement in PHENIX for cryo-EM and crystallography. Acta crystallographica Section D, Struct. Biol. **74**(Pt 6) (2018). https://doi.org/10.1107/S2059798318006551
6. Nguyen, T., Wriggers, W., He, J.: A Data Set of Paired Structural Segments Between Protein Data Bank and AlphaFold DB for Medium-Resolution Cryo-EM Density Maps: A Gap in Overall Structural Quality. In: Peng, W., Cai, Z., Skums, P. (eds.) Bioinformatics Research and Applications, pp. 52–63. Springer Nature Singapore, Singapore (2024)
7. Jumper, J., Evans, R., Pritzel, A., Green, T., Figurnov, M., Ronneberger, O., et al.: Highly accurate protein structure prediction with AlphaFold. Nature **596**(7873), 583–589 (2021). https://doi.org/10.1038/s41586-021-03819-2
8. Kryshtafovych, A., Schwede, T., Topf, M., Fidelis, K., Moult, J.: Critical assessment of methods of protein structure prediction (CASP)-Round XIV. Proteins **89**(12), 1607–1617 (2021). https://doi.org/10.1002/prot.26237
9. He, J., Lin, P., Chen, J., Cao, H., Huang, S.-Y.: Model building of protein complexes from intermediate-resolution cryo-EM maps with deep learning-guided automatic assembly. Nat. Commun. **13**(1), 4066 (2022). https://doi.org/10.1038/s41467-022-31748-9
10. Chen, J., Zia, A., Luo, A., Meng, H., Wang, F., Hou, J., et al.: Enhancing cryo-EM structure prediction with DeepTracer and AlphaFold2 integration. Briefings Bioinf. **25**(3), bbae118 (2024). https://doi.org/10.1093/bib/bbae118
11. Abramson, J., Adler, J., Dunger, J., Evans, R., Green, T., Pritzel, A., et al.: Accurate structure prediction of biomolecular interactions with AlphaFold 3. Nature **630**(8016), 493–500 (2024). https://doi.org/10.1038/s41586-024-07487-w
12. The ww PDBC. EMDB—the electron microscopy data bank. Nucleic Acids Res. **52**(D1), D456–D65 (2024). https://doi.org/10.1093/nar/gkad1019
13. Liebschner, D., Afonine, P.V., Baker, M.L., Bunkóczi, G., Chen, V.B., Croll, T.I., et al.: Macromolecular structure determination using X-rays, neutrons and electrons: recent developments in Phenix. Acta Crystallographica Section D: Struct. Biol. **75**(10), 861–877 (2019)
14. Mukherjee, S., Zhang, Y.: MM-align: a quick algorithm for aligning multiple-chain protein complex structures using iterative dynamic programming. Nucleic Acids Res. **37**(11), e83 (2009). https://doi.org/10.1093/nar/gkp318
15. Zhang, Y., Skolnick, J.: TM-align: a protein structure alignment algorithm based on the TM-score. Nucleic Acids Res. **33**(7), 2302–2309 (2005). https://doi.org/10.1093/nar/gki524

16. Wriggers, W.: Conventions and workflows for using Situs. Acta Crystallogr. D Biol. Crystallogr. **68**(Pt 4), 344–351 (2012). https://doi.org/10.1107/s0907444911049791
17. Pettersen, E.F., Goddard, T.D., Huang, C.C., Meng, E.C., Couch, G.S., Croll, T.I., et al.: UCSF ChimeraX: Structure visualization for researchers, educators, and developers. Protein Sci. **30**(1), 70–82 (2021)
18. Mu, Y., Sazzed, S., Alshammari, M., Sun, J., He, J.: A tool for segmentation of secondary structures in 3D cryo-EM density map components using deep convolutional neural networks. Front. Bioinf. **1**, 51 (2021)
19. Mu, Y., Nguyen, T., Hawickhorst, B., Wriggers, W., Sun, J., He, J.: The combined focal loss and dice loss function improves the segmentation of beta-sheets in medium-resolution cryo-electron-microscopy density maps. Bioinform. Adv. **4**(1), vbae169 (2024). https://doi.org/10.1093/bioadv/vbae169

Automatic Explanation of Protein-Protein Binding Mechanism: A Preliminary Study

Justin Z. Tam[1], Yangying Liu[1], Dhruv S. Jain[2], Grant Armstrong[1], and Brian Y. Chen[1]

[1] Lehigh University, Department Computer Science and Engineering, Bethlehem, PA, USA
chen@cse.lehigh.edu

[2] Indian Institute of Technology Bombay, Department Computer Science and Engineering, Mumbai, Maharashtra, India

Abstract. Understanding the biochemical mechanisms that drive protein-protein interactions is a challenging task, traditionally requiring mutation studies and expert interpretation of protein structures. A method that can generate mechanistic explanations from the biochemical properties and contributions of interactions would enhance our ability to study protein-protein interactions. In this study, we present a novel approach to interpreting mechanistic insights from machine learning methods; we manually annotated a dataset of 1225 mutation experiments with mechanistic insights focused on electrostatic, hydrogen bonding, steric and hydrophobic interactions. To show a preliminary process for evaluating mechanism prediction models, we extracted SHAP features that are representative of protein binding mechanisms from a Gradient-Boosting Tree (GBT) model trained to predict binding affinity. We found that the SHAP values generally agreed with the annotated mechanisms from our dataset, especially when looking at electrostatic and steric features. We also found that hydrophobicity consistently played a dominant role and hydrogen bonds consistently played a secondary role, challenging conventional assumptions about the role of these interactions.

Keywords: Protein Structure/Function · Protein-protein Interactions

1 Introduction

Interacting proteins play a central role in organizing the functions of the cell. They form complexes that regulate the cell cycle [9], present antigens for the adaptive immune system [4], influence the plasticity of the nervous system [21], and many other important tasks. In order to reverse engineer these interactions for the study of healthy systems, or to create interventions for unhealthy ones,

---

J. Z. Tam and Y. Liu—equal contribution.

© The Author(s), under exclusive license to Springer Nature Switzerland AG 2025
N. Haspel and K. Molloy (Eds.): CSBW 2024, CCIS 2396, pp. 84–97, 2025.
https://doi.org/10.1007/978-3-031-85435-4_7

it is essential to discover the biochemical mechanisms that underpin selective binding. To this end, the traditional approach has been to perform mutations of amino acids at the interface and infer the role they play in binding. This process is expensive and time consuming, because there are many amino acids involved and multiple chemical interactions to consider.

To hasten this approach, categories of software are used to help select amino acids for mutation. Structure prediction (e.g. [1,2]) and docking algorithms (e.g. [10,13]) predict the structure of proteins in complex in order to illustrate which amino acids are present at the interface and what interactions they form. When using these tools, researchers visually examine the predicted complex and imagine how the biochemical mechanisms interact, building insight into the how the binding mechanism might work. Molecular simulation tools (e.g. [5,7]) form a second category of software that offer insights into how protein structures change in order to recognize each other. Careful scrutiny of simulated motions offers insight into how specific amino acids interact with each other and a second distinct perspective on the possible roles they play in binding.

These are powerful methods, but they rely substantially on the capacity of the human user to integrate their outputs into models of interacting biochemical mechanisms. When there are many variations, or many mechanisms involved, the number of possibilities can easily exceed human attention. To examine this problem, this paper explores the possibility of creating structural analyses that infer the mechanistic role of amino acids directly. Our approach is to interpret the features of a machine learning method that was originally designed to predict binding affinity between two proteins. Rather than examining the accuracy of the binding affinity predictions, which we do elsewhere [19], we asked whether or not accurate explanations could be generated for the biochemical contributions of specific amino acids.

Evaluating this question requires a ground truth dataset that describes the biochemical role of individual amino acids and maps it to a discrete list of mechanism. This could simply be a list of tags that indicate, for example, that amino acid X plays an electrostatic role in binding, or that amino acid Y forms hydrogen bonds to support binding. While the diversity of structural biology will always exhibit novel mechanisms for molecular function, these basic mechanisms can form the building blocks larger interactions. This information is common in the structural literature, but it exists as freeform text and is not organized as a ground truth dataset that can be leveraged for explanations.

For this reason, one major contribution of this paper is to begin creating that dataset. Our dataset was constructed from a subset of the SKEMPI 2.0 database [11], which connects the structures of experimentally determined protein complexes with the binding affinity of different mutations of amino acids in the complex [11]. We manually examined the literature relating to these structures, identifying specific quotes that document the mechanisms in which many amino acids are involved. In this work, we tagged mechanisms that are documented to have [steric], [electrostatic], [hydropathic], [salt bridge] or [hydrogen bonding] contributions to binding affinity, and we denote them with square brackets. This

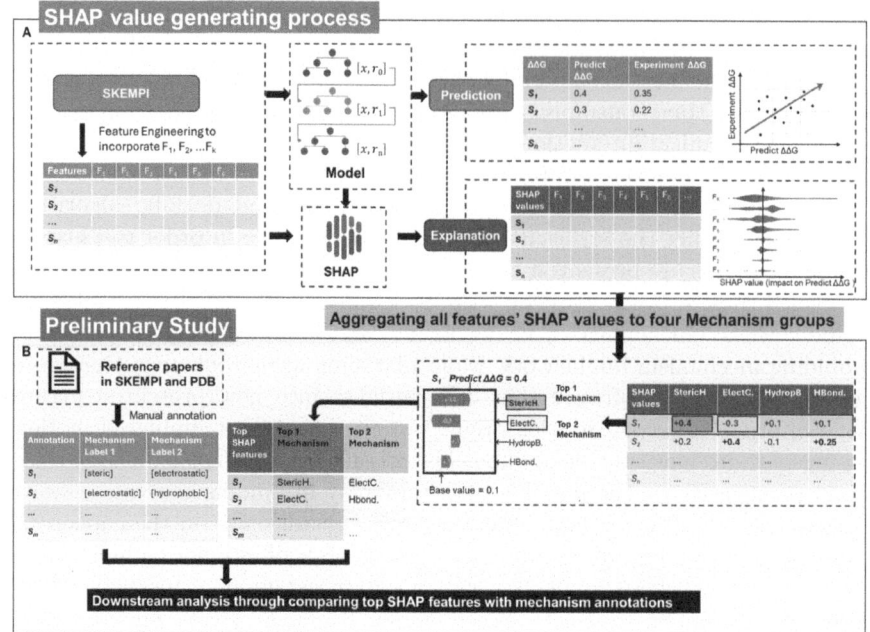

Fig. 1. Overview of the Preliminary Study and SHAP Value Generation Process. A) SHAP Value Generation Workflow: SHAP is applied alongside a binding affinity prediction model. The prediction model computes changes in binding affinity, while SHAP provides biochemical explanations underlying these changes for the SKEMPI data. B) Preliminary Study Workflow: Building on SHAP values obtained from a previous publication, we aggregated all feature SHAP values into four mechanism types and identified the top 1 and top 2 mechanisms for each sample. We then conducted a comprehensive literature review of the SKEMPI dataset and manually curated a PPI mutation mechanism annotation dataset. Finally, we compared the mechanisms derived from the model-generated explanations (top SHAP features) with the annotated mechanism labels from literature sources to assess the consistency and accuracy of the model's biochemical interpretations.

dataset, while already sizable, remains incomplete, as we continue to add additional literature and additional mechanisms. Currently, we have annotated 1225 experiments from the SKEMPI dataset and aim to complete 2639 annotations in total.

Our explanations are generated using SHAP (SHapley Additive exPlanations) values [20], which offer a way to explain the importance of different feature groups used by a Gradient-Boosting Tree (GBT) algorithm. A number of GBT-based algorithms are compatible with this purpose, including topNetTree [22], GeoPPI [18], MechPPI [19], and MuToN [17]. In this work, we use MechPPI, and we generate SHAP values from steric, electrostatic, hydropathic, and hydrogen bonding feature groups computed for a mutant. We treat high absolute

SHAP values for a given feature group (e.g. the electrostatic group of features) as being an indicator that the mutated amino acid influences affinity through that group's mechanism (e.g. an electrostatic influence). Measuring the correlation between SHAP values and our literature annotations, we observed numerous correlations, indicating that this strategy for explaining binding mechanisms can be effective, based on our well defined ground truth data (Fig. 1).

Our method is the first technique, to our knowledge, for automatically generating explanations of the biochemical role of specific amino acids in protein-protein interfaces. In our results, we observed varying degrees of accuracy in identifying mechanisms of different types, and correlations between some mechanisms that have similar physical foundations. Two examples from our detailed literature search illustrate these relationships. Overall, these findings suggest that structure based features can indeed be used to generate explanations about the roles of amino acids.

2 Methods

2.1 Manual Annotation Process

The labeling of SKEMPI mutation mechanisms was carried out through a manual, multi-step process to ensure high-quality and reliable mechanism information. We first verified that the mutant in SKEMPI was accurately referenced in the literature. The authors then reviewed the papers that were referenced by SKEMPI and the PDB [3], seeking mechanistic information about the amino acid in question. Generally, the amino acid that was mutated because it's biochemical role was being established experimentally. The authors then annotated the role of the amino acid with tags representing biochemical interactions, which in this work are [hbond], [electrostatic], [salt bridge], [steric], or [hydrophobic]. These labels for experimentally documented interactions are indicated using square brackets (e.g., [hbond], [electrostatic]) throughout this paper, so that they cannot be easily confused with mechanistic predictions with similar names. Note that we have distinct annotations for [salt bridge] and [electrostatic] mechanisms despite their electrostatic similarity; we define them separately due to the notable differences in effect between short range salt bridge interactions and general long range ionic interactions [15]. For each entry, the final mechanism annotation includes: PDB structure, mutated amino acid, mechanism tag, relevant quotes taken directly from the referenced literature, as well as the paper reference for traceability. While this process is extremely time consuming, it ensures high quality annotations with well defined mechanistic information.

2.2 SHapley Additive ExPlanations (SHAP Values)

Prediction models often use numerous features, each contributing towards the final output prediction. However, the degree of contribution from each feature can be difficult to decipher from the final prediction. To quantify these contributions, we used SHapley Additive exPlanations (SHAP), a method that assigns an importance value to each feature based on its contribution on the model's prediction. These values were extracted from a pre-trained regression model which predicts binding affinity of protein complexes [19]. To determine the importance of a feature j, we calculate the marginal contribution - the difference in the model's prediction when the feature is included versus when it is excluded.

The marginal contribution of feature j depends on the subset of features to which j is added (subsets of size 1, 2, 3, ..., up to N). SHAP values are calculated by determining all possible combinations of feature subsets, computing the total marginal contribution of feature j for each subset, and finding the average of these contributions. The average contribution is weighted based on the number of subset permutations, ensuring equal distribution of feature contributions across all possible combinations. This results in a SHAP value that represents the average contribution of a feature across all possible combinations. The SHAP value calculation can be formally expressed as:

$$\phi_i = \sum_{S \subseteq N \setminus \{i\}} \frac{|S|!(N - |S| - 1)!}{N!} [f_x(S \cup \{i\}) - f_x(S)] \qquad (1)$$

Here, ϕ_i is the SHAP value for feature i, S is the subset of features from all features N except for features i, N is the total number of features, for each feature i, we need to consider all possible subsets S of features that do not include i ($S \subseteq N \setminus \{i\}$), $\frac{|S|!(N-|S|-1)!}{N!}$ is the weighting factor counting the number of permutations of the subset S, $f_x(S)$ is the prediction when only the features in subset S are present, $f_x(S \cup \{i\})$ is the prediction when feature i is added to subset S, marginal contribution of feature i is $f_x(S \cup \{i\}) - f_x(S)$, it represents the change in the prediction when feature i is included in the set compared to when it is excluded. So the SHAP value ϕ_i for feature i is the sum of its weighted marginal contributions across all subsets.

After obtaining SHAP values of all features, we can use SHAP values to explore the most influential mechanisms from a mutation on the binding affinity of protein complexes. The larger the absolute SHAP value a feature has, the greater the contribution that feature has towards the final prediction.

2.3 Interactions Represented by SHAP Values

In this work, we divide the features of MechPPI into four groups: electrostatic complementarity, hydrogen bonding, steric interactions, and hydrophobic interactions. One SHAP value was computed for each group. Below, we describe some specific details also covered by SHAP values within each interaction and specify the number of SHAP features assigned to each group.

For our electrostatic feature group, our GBT method uses VASP-E to measures electrostatic complementarity by using Constructive Solid Geometry (CSG) operations to generate solid representations of electrostatic fields [8]. This approach generates regions of electrostatic complementarity by computing the overlapping volume between multiple positive and negative isopotentials from interacting molecules. This method is particularly useful in finding bond interactions like salt bridges which innately involve oppositely charged residues.

For our hydrogen bonding feature group, we identify the gain and loss of hydrogen bonds at the binding site and at the binding interface. We count the number of backbone-backbone, backbone-sidechain, sidechain-sidechain, and sidechain-backbone hydrogen bonds. Finally, we use a volumetric analysis that represents the range of bond angles and bond distances accessible to a hydrogen bond [6].

We also have a steric feature group that uses a solid volumetric representation of the interface molecular surface in a protein interface. It also incorporates the volume of the mutated residue and distance of mutated residue to the binding partner. The SHAP value that represents this group helps capture the contribution of steric hindrance and Van der Waals complementarity towards protein-protein interactions.

Hydrophobic interactions are known to provide substantial stability for protein binding [12]. In our hydrophobic feature group, we describe hydropathy environments using counts of neighboring hydrophobic, hydrophilic, and neutral residues to the mutant residue.

2.4 Evaluating SHAP Feature Contribution to Mechanisms

We developed a systematic process for selecting the top-contributing mechanism group to a protein-protein interaction based on SHAP values, and use these to compare contributions of each mechanism group to the recorded mechanistic interaction observed in SKEMPI proteins.

Aggregating SHAP Features. The 97 SHAP features i were divided into four categories based on how each SHAP feature is representative of the physicochemical interaction mechanisms discussed above, which we call mechanism groups. The SHAP values associated to each SHAP feature are then aggregated by mechanism group by taking the absolute sum A; this is done separately for each mutation. This ensures that both positive and negative contributions are captured equally, reflecting the magnitude of impact of all SHAP values for each mechanism group.

$$A_{mut} = \sum_{i \in X} |\phi_i|, \quad \text{for } X \in \{E, H, S, P\} \tag{2}$$

Counting Top Mechanism Groups. For each mutation, we calculated an aggregate SHAP value A_X for each major mechanism group. We used these aggregate SHAP values to produce two mechanism prediction strategies. The first strategy finds the mechanism group that has the highest aggregate SHAP value; we predict that the mechanism of the mutated amino acid is that of the mechanism group that has the highest absolute SHAP value and we call this the "top 1" mechanism G_{top1}. For example, if, for a given amino acid, the SHAP values associated with electrostatic complementarity has the highest aggregate SHAP value, then the "top 1" prediction strategy predicts that the amino acid in question influences binding via an electrostatic complementarity mechanism.

$$G_{top1} = \arg \max_{X \in \{E,H,S,P\}} A_{mut} \quad (3)$$

Our second strategy makes the same prediction as "top 1" and also finds the mechanism group with the second highest SHAP value; the mechanism group of both the highest and second highest SHAP value we call the "top 2" mechanisms. With the "top 2" mechanisms, we predict that the primary and secondary mechanism of the mutated amino acid is that of the "top 2" mechanism groups. In this scenario, it might predict, for example, that the amino acid acts via an electrostatic mechanism as the top 1 primary mechanism and hydrogen bonding as the top 2 secondary mechanism.

In our results, we illustrate how often these predictions coincide with tagged annotations from SKEMPI. The five tagged annotations are [electrostatic], [hydrogen bond], [steric], [hydrophobic], and [salt bridge]. We specifically evaluate salt bridge annotations because of its high degree of strength compared to long-range electrostatic interactions, its innate electrostatic properties, as well as its co-occurrence with hydrogen bonds and electrostatic interactions. As a result of these steps, we obtain a count of how many times each major mechanism group had the top 1 or top 2 SHAP value across the subset of mutations labelled with the five interaction mechanisms mentioned. The count is divided by the total number of mutations for each mechanism subset and is presented in the results.

3 Results

3.1 Annotating Mechanisms from SKEMPI

The SKEMPI dataset maps mutations in protein-protein interactions (PPIs) to changes in binding free energy, relative to a wild type complex. While SKEMPI provides extensive mutation data, it lacks detailed mechanistic information on the biochemical interaction that occurs. To bridge this gap, we undertook a manual annotation process to supplement the dataset with mechanistic insights.

First, we identified 1225 distinct mutation experiments from 121 proteins described in the SKEMPI dataset. For each amino acid mentioned in SKEMPI, we manually searched through the literature references by SKEMPI and the Protein Data Bank, identifying any specific interactions related to the amino acid that could explain how the mutant affects the protein-protein interaction.

From the 1225 mutation experiments, we removed several proteins where mechanistic information are difficult to extract. 374 mutations had no discussion of mechanistic information about the amino acid. The majority of these mutations came from alanine-scanning experiments which did not consider the structural role of individual amino acids. Another 39 un-annotated mutations were from T-cell receptors (TCR) interacting with a peptide in the major histocompatibility complex (pMHC). We found that most papers focused their discussion on the general TCR/pMHC/ligand complex rather than individual amino acids and interactions, and so we removed all TCR/pMHC mutations. When looking at just the five mechanisms tagged in this work, our manual annotation effort resulted in a total of 819 amino acids with documented mechanisms. The composition of these annotations is shown in Table 1.

Table 1. Count of occurrences for each biochemical mechanism.

Mechanism	hbond	hydrophobic	salt bridge	electrostatic	steric
Count	470	139	101	79	30

Note that this dataset of biochemical mechanisms is still a work in progress. Our subset consists of a sequence non-redundant selection of the SKEMPI dataset, reducing biases from highly similar sequences and ensuring diversity in the biochemical interactions studied, but quite a few proteins have not yet been annotated. Nonetheless, the number of annotations is not small, so we feel that this initial study could provide some insight into the value of producing the whole dataset.

Altogether, we annotated mechanisms for 819 mutants and collected the SHAP values for those same mutants to produce explanations of their binding mechanisms.

3.2 Comparing SHAP Features with Mechanism Annotations

Here, we evaluate how well the SHAP values from our dataset predict annotated binding mechanisms. The SHAP values were aggregated into four major mechanism groups: electrostatic, hydrogen bond, steric, and hydrophobic interactions. Next, we performed both the top 1 or top 2 prediction strategy for each mutation in our SKEMPI subset. These selections were used to count the proportion of the top contributing mechanism and evaluated if mechanism annotations of the SKEMPI data agreed with the top contributing mechanism.

Two sets of results are presented in Fig. 2. The top 1 predictions are shown in the bottom segment of each bar (dark colors), while the top 2 predictions are shown in a long bar behind them (light colors). Since top 2 predictions strictly contain all predictions made by the top 1 strategy, the top 2 bar must always be at least as tall as the top 1 bar. Each bar is presented as a proportion (left axis) of the dataset.

Top 1 Mechanism Group Analysis. The bottom segments (dark colors) in Fig. 2 show the proportion of how frequently a mechanism group has the highest SHAP value G_{top1}, and therefore plays the most dominant role in prediction for each annotated mechanism tag. For example, when observing [hbond] labelled mutations (dark blue) in the "Electrostatic" mechanism group (leftmost column grouping), we see that for mutations tagged as involving hydrogen bonding, around 27% of these mutations have electrostatic as the dominant mechanism group based on SHAP value. In comparison, around 58% of mutations tagged as involving [salt bridges] (dark green) have electrostatic as the dominant mechanism group. This is consistent with the involvement of electrostatic interactions in salt bridge formation, as opposed to hydrogen bonds which involve only partial charges.

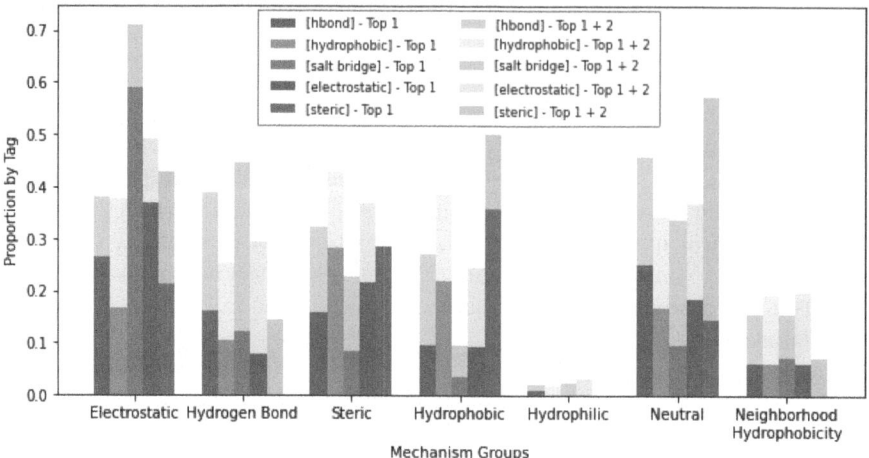

Fig. 2. The proportion of top 2 counts for each mechanism tag, showing the combined contribution of both dominant and secondary mechanisms. Dark color bars can be interpreted as the proportion of annotated mutations which have the mechanism group as the top 1 SHAP value. Light color bars are the proportion of annotated mutations which have the mechanism group as the top 2 SHAP value.

In general, we observe substantial agreement between SHAP mechanism groups and our mechanism annotations. For [salt bridge] and [electrostatic] tagged mutations, the electrostatic mechanism group is most frequently the dominant mechanism, consistent with expectations. Similarly, [steric] and [hydrophobic] tagged mutations are well-aligned with the steric and hydrophobic mechanism groups. Surprisingly, the hydrogen bond mechanism group appears less frequently as the top 1 feature for [hbond] tagged mutations, where the electrostatic mechanism was more prevalent than expected.

Top 2 SHAP Feature Analysis. The lighter colored columns in Fig. 2 show the proportion of major mechanism group contributions when the top 2 mechanism groups G_{top2} within each category are selected for each mutation. This figure provides insight into the secondary contributions of mechanism groups, complementing the top 1 analysis. In general, the top 2 mechanisms show slightly better alignment with our expectation for the relationship between SHAP values and annotated mechanism tags than for top 1. For example, hydrogen bonds are more frequently a top 2 contributing mechanism for [hbond] tagged mutations, as opposed to the top 1 where hydrogen bonds were significantly lower in proportion than electrostatic for [hbond] tagged mutations; similarly, the SHAP value for the steric mechanism group are better represented for [steric] tags.

The electrostatic mechanism group continues to play a significant role for mutations labeled as involving salt bridges, with more than 70% of these mutations having electrostatics among the top 2 mechanism groups. This reinforces the earlier observation of electrostatics in salt bridge formation. Interestingly, hydrogen bonding and "hydroneutral" mechanism contributions had the greatest increase across most tags in the top 2 analysis, suggesting a more subtle role for hydrogen bonding and neutral hydrophilicity that may not appear when only considering the top 1 mechanism.

In general, the inclusion of the top 2 analysis highlights the importance of considering multiple mechanisms in understanding complex mutations, as interactions that were previously less visible (e.g., steric and hydrogen bonding) now show greater influence.

Despite several trends aligning with known biochemical principles, some results in Fig. 2 do not match expectations. Notably, the hydrophilic mechanism group had almost no frequency as top 1 or top 2; this is in stark contrast to the relatively large proportion for hydrophobic and neutral mechanism groups across all tags. Additionally, SHAP values for hydrogen bonding remain lower than neutral hydrophilicity for [hbond] labels, suggesting that hydrogen bond SHAP features may not be the main contributing force in predicting [hbond] tagged mutations.

Examples of Top SHAP Features Supported by Other Studies. Here we present two examples that illustrate the explanations generated by our method. The first example is derived from the complex of RAP and C-RAF1 (pdb: 1C1Y) [23], where ARG67 was mutated to Ala. The mutation R67A is shown in Fig. 3 where the mutation of the original Arg residue creates a reduction in electrostatic isopotential and therefore a reduction in electrostatic complementarity. This shows that the electrostatic mechanism is the dominant mechanism underlying the mutation R67A. This is consistent with the specific top SHAP feature found for this sample. The prediction is verified in the literature from Zhang, et al. [23], who state: *"Similar to the RalGDS complexes, the polar and positively charged amino acid side-chains of the effector contribute most to the interaction, namely Arg59, Arg67, Lys84 and Arg89..."*.

The second example is based on the complex of the BMP-2 Receptor and the N-Terminal Von Willebrand factor type C domain of crossveinless 2 (CV-2) (pdb: 3BK3). This complex is involved in Bone Morphogenetic Protein (BMP) signaling pathways [14]. We show an example of mutation of Ile21 to Ala in Fig. 3. The top SHAP feature we generated for this mutation is hydrophobic

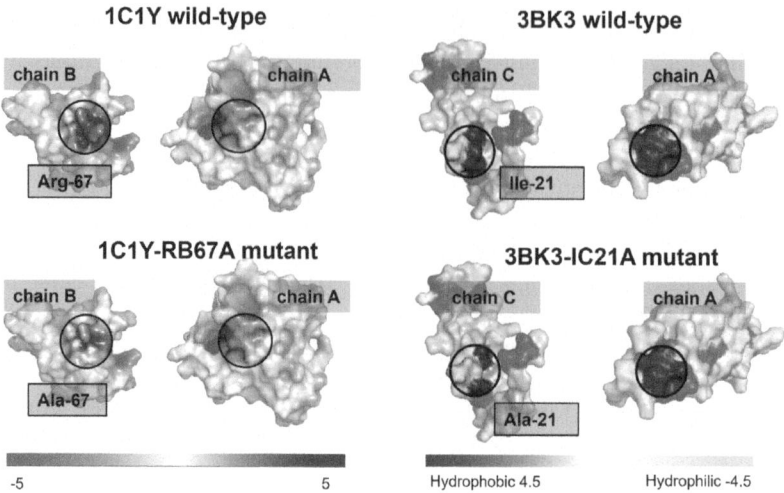

Fig. 3. Examples of two protein complexes 1C1Y and 3BK3, their Top 1 mechanisms underlying mutation of 1C1Y-RB67A and 3BK3-IC21A. Both complexes are illustrated like an open book, where the interfacial regions are opened from the center to face the reader. The regions of the binding interface that are nearest to the mutated amino acid are circled. A red/blue scale of surface charge is shown from -5 kt/e to $+5$kt/e. For the 1C1Y-RB67A mutation, electrostatic complementarity reduction is the predicted mechanism of the RB67A mutation (according to the top 1 SHAP feature). This complementarity reduction is apparent in the reduction of blue surface area on chain B in the mutant. For 3BK3-1C21A, reduction of hydrophobic contact area induced by mutation is predicted to be the most influential mechanism (according to the top 1 SHAP feature), and is visible in the reduction of purple surface area on chain C in the mutant. A Kyte-Doolitle hydrophobicity scale [16] is provided from purple (hydrophobic) to yellow (hydrophillic). These two predicted mechanisms are consistent with earlier structural studies [14, 23]. (Color figure online)

which is expected when mutating from a large hydrophobic residue like Ile to a smaller, less hydrophobic Alanine as well as a reduction in hydrophobic contact area. Therefore, this indicates hydrophobic mechanism is the most important mechanism underlying mutation I21A, which is also consistent with our predicted top SHAP feature. This observation is verified in the literature by Kiel, et al. [14], who write *"...large hydrophobic side chains (mostly tryptophane) at a position corresponding to Ile21...Upon mutation of Ile21 or Ile27 to Ala, about 1.3 kcal/mol of binding free energy is lost."* Together, these examples illustrate how the predictions made with our tool and can be used to produce mechanistic explanations that are consistent with quotes structural publications.

4 Discussion

We present our work which uses explainable products of a trained prediction model to suggest a biochemical mechanism for amino acids involved in protein-protein interactions. Specifically, we evaluated the alignment between SHAP values from a GBT model trained to predict binding affinity and ground truth mechanistic annotations from experimental data. This preliminary work introduces a process for integrating fine-grained explainable model outcomes directly with functional real-world lab reports, allowing us to evaluate specific mechanisms while bypassing the need for exhaustive prediction processes typically employed to approximate real-world data.

In order to develop a ground truth mechanistic dataset, we used a novel annotation process to associate experimental mutation data from the SKEMPI 2.0 dataset with relevant mechanism tags. On SKEMPI dataset annotation, we determined the mechanism of 1225 amino acid mutation experiments through intensive manual literature review. Our goal in annotating mechanisms directly is to use these annotations as labels or features in downstream prediction tasks. In this study, we evaluated the contribution of four different interaction mechanisms -electrostatic, hydrogen bonding, steric, and 4 categories of hydrophobic interactions - by comparing SHAP values generated from a GBT model with the annotated mechanism tags.

To allow comparison, we grouped our SHAP features into 7 "mechanism groups" - electrostatic, hydrogen bonding, steric, and 4 different hydrophobicity groups - and evaluated which of the mechanism groups have the top 1 and top 2 total SHAP value among the four. Across the board, we found substantial agreement between SHAP mechanism groups and our annotated mechanisms for both top 1 and top 2 mechanisms. We notice, however, that hydrophobic and neutral mechanisms are much more prominent for many different tagged mutations, like [steric], [hydrophobic] and [hbond] tagged mutations. This suggests that hydrophobicity may play a more important role in stabilizing these interactions than previously recognized. Electrostatic interactions, as expected, were dominant for [electrostatics] tagged mutations and particularly dominant in [salt bridge] tagged mutations, confirming their crucial role in salt bridge formation. Unexpectedly, hydrogen bond SHAP features have low contribution across all tags in Top 1, especially for [hbond] tagged mutations where we expected for hydrogen bond mechanism to be the highest contributing SHAP value. Similarly, steric contributions were lower than anticipated across for all tags except [hydrophobic] tag, indicating that their influence on binding affinity may be hidden by higher contributing forces like hydrophobicity or is less effectively captured in our model.

In the top 2 mechanism analysis, the contribution of secondary interactions became clearer. The SHAP value for hydrogen bonding mechanism group, for example, showed a more significant role in [hbond] tagged mutations; similarly, steric and hydroneutral interactions were better represented. This highlights the importance of considering multiple features when evaluating complex mutations, as interactions that are hidden by the top contributing mechanism can still play an essential and complementary role. Overall, the inclusion of both top 1 and top 2 analyses generally agreed with the mechanism tags while providing more insights to how primary and secondary interactions cooperate in protein-protein interactions.

Despite the higher contribution of hydrogen bond and steric to [hbond] and [steric] tagged mutations respectively when examining top 2 mechanisms, these two mechanism groups still differ from classical expectations. Models, especially in energy function models and in traditional protein structure models, have assumed that hydrogen bonds and steric effects are significant to protein-protein interactions, however our results challenge these traditional assumptions. Overall, our results suggest that SHAP values can effectively capture both dominant and subtle mechanistic influences on binding affinity. In addition, we found that our manual SKEMPI annotations may work well with protein interaction models that utilize fine-grained mechanistic features or with explainable values like SHAP.

These differences also indicate the presence of additional structural nuances in this approach to producing biochemical explanations for binding mechanisms. Where explanations based on features disagree with annotations, it is important to understand how such disagreements can occur, and whether it results from a matter of biochemical representations, a lack of detailed annotations, or both.

One way to better capture interaction mechanisms is to review and completely annotate all interactions in the SKEMPI 2.0 dataset. This would result in a curated mechanism annotation dataset of over 2600 mutation experiments. We established a procedure for labeling and tagging large interaction datasets for mechanisms; the resulting dataset would be useful in many downstream tasks that are impossible today due to a lack of datasets providing biochemical interaction insights. For example, this dataset could be augmented for language-based tasks with recent advancements in LLM-based systems. Altogether, the practice of explaining how protein structures function, based on structural data working in tandem with annotations derived from the literature, represents a rich opportunity to advance the field.

Data. We have made the mechanism annotation data subset of SKEMPI available here.

Acknowledgements. This work was funded in part by NIH grant R01GM123131 to Brian Y. Chen.

References

1. Abramson, J., et al.: Accurate structure prediction of biomolecular interactions with AlphaFold 3. Nature 1–3 (2024)
2. Baek, M., et al.: Accurate prediction of protein structures and interactions using a three-track neural network. Science **373**(6557), 871–876 (2021)
3. Berman, H.M., et al.: The protein data bank. Nucleic Acids Res. **28**(1), 235–242 (2000)
4. Bjorkman, P.J., Parham, P.: Structure, function, and diversity of class i major histocompatibility complex molecules. Annu. Rev. Biochem. **59**(1), 253–288 (1990)

5. Brooks, B.R., et al.: CHARMM: the biomolecular simulation program. J. Comput. Chem. **30**(10), 1545–1614 (2009)
6. Buranasilp, C., Chen, B.Y.: A conical representation of hydrogen bond geometry for quantifying bond interactions. In: 2021 IEEE International Conference on Bioinformatics and Biomedicine (BIBM), pp. 2479–2486. IEEE (2021)
7. Case, D.A., et al.: Amber 2024. University of California, San Francisco (2024)
8. Chen, B.Y.: VASP-E: specificity annotation with a volumetric analysis of electrostatic Isopotentials. PLoS Comput. Biol. **10**(8), e1003792 (2014)
9. Hartwell, L.H., Weinert, T.A.: Checkpoints: controls that ensure the order of cell cycle events. Science **246**(4930), 629–634 (1989)
10. Honorato, R.V., et al.: The HADDOCK2. 4 web server for integrative modeling of biomolecular complexes. Nat. Protoc. 1–23 (2024)
11. Jankauskaitė, J., Jiménez-García, B., Dapkūnas, J., Fernández-Recio, J., Moal, I.H.: SKEMPI 2.0: an updated benchmark of changes in protein–protein binding energy, kinetics and thermodynamics upon mutation. Bioinformatics **35**(3), 462–469 (2019)
12. Jernigan, R.L., Khade, P., Kumar, A., Kloczkowski, A.: Using surface hydrophobicity together with empirical potentials to identify protein–protein binding sites: application to the interactions of e-cadherins. In: Computer Simulations of Aggregation of Proteins and Peptides, pp. 41–50. Springer (2022)
13. Jiménez-García, B., Pons, C., Fernández-Recio, J.: pyDockWEB: a web server for rigid-body protein-protein docking using electrostatics and desolvation scoring. Bioinformatics **29**(13), 1698–1699 (2013)
14. Kiel, C., Serrano, L., Herrmann, C.: A detailed thermodynamic analysis of ras/effector complex interfaces. J. Mol. Biol. **340**(5), 1039–1058 (2004)
15. Kurczab, R., Śliwa, P., Rataj, K., Kafel, R., Bojarski, A.J.: Salt bridge in ligand-protein complexes–systematic theoretical and statistical investigations. J. Chem. Inf. Model. **58**(11), 2224–2238 (2018)
16. Kyte, J., Doolittle, R.F.: A simple method for displaying the hydropathic character of a protein. J. Mol. Biol. **157**(1), 105–132 (1982)
17. Li, P., Liu, Z.P.: Muton quantifies binding affinity changes upon protein mutations by geometric deep learning. Adv. Sci. 2402918 (2024)
18. Liu, X., Luo, Y., Li, P., Song, S., Peng, J.: Deep geometric representations for modeling effects of mutations on protein-protein binding affinity. PLoS Comput. Biol. **17**(8), e1009284 (2021)
19. Liu, Y., Armstrong, G., Tam, J., Chen, B.Y.: MechPPI: binding mechanism-based machine-learning tool for predicting protein-protein binding affinity changes upon mutations. bioRxiv, pp. 2023–10 (2023)
20. Lundberg, S.: A unified approach to interpreting model predictions. arXiv preprint arXiv:1705.07874 (2017)
21. Miwa, J.M., Lester, H.A., Walz, A.: Optimizing cholinergic tone through lynx modulators of nicotinic receptors: implications for plasticity and nicotine addiction. Physiology **27**(4), 187–199 (2012)
22. Wang, M., Cang, Z., Wei, G.W.: A topology-based network tree for the prediction of protein-protein binding affinity changes following mutation. Nat. Mach. Intell. **2**(2), 116–123 (2020)
23. Zhang, J.l., et al.: Crystal structure analysis reveals how the chordin family member crossveinless 2 blocks BMP-2 receptor binding. Dev. Cell **14**(5), 739–750 (2008)

Author Index

A
Afrasiabi, Fatemeh 45
Alam, Fardina Fathmiul 1
Alshammari, Maytha 31
Armstrong, Grant 84

C
Chen, Brian Y. 84
Christensen, Katie 58
Coffland, Sarah 58

D
Dehghanpoor, Ramin 45

F
Forouzesh, Negin 16

H
Haspel, Nurit 45
He, Jing 31, 71

J
Jagodzinski, Filip 58
Jain, Dhruv S. 84
Jasko, Ari 16

K
Kanani, Riya 1

L
Li, Changrui 58, 71
Liu, Yangying 84

N
Nguyen, Thu 71

P
Patre, Shraddha 1

S
Sagar, Dikshant 16

T
Tam, Justin Z. 84

W
Wriggers, Willy 31, 71

The manufacturer's authorised representative in the EU is Springer Nature Customer Service Centre GmbH, Europaplatz 3, 69115 Heidelberg, Germany. If you have any concerns regarding our products, please contact ProductSafety@springernature.com

Printed and bound by CPI Group (UK) Ltd, Croydon, CR0 4YY

26/03/2026

02078963-0006